Update on Malacology

Edited by Sajal Ray
and Soumalya Mukherjee

Published in London, United Kingdom

IntechOpen

Supporting open minds since 2005

Update on Malacology
http://dx.doi.org/10.5772/intechopen.94710
Edited by Sajal Ray and Soumalya Mukherjee

Contributors
Latife Ceyda İrkin, Minoru Saito, Yoshimasa Komatsuzaki, Ayaka Itoh, M.P. Charó, Sidy Bakhoum, Christopher J.E. Haggerty, Cheikh Tidiane Ba, Gilles Riveau, Nicolas Jouanard, Jason Robert Rohr, Shailesh Saurabh, Sweta Pradhan, Sonal Suman, Wolyu Korma Erkano

First published in London, United Kingdom, 2022 by IntechOpen
IntechOpen is the global imprint of INTECHOPEN LIMITED, registered in England and Wales, registration number: 11086078, 5 Princes Gate Court, London, SW7 2QJ, United Kingdom
Printed in Croatia

British Library Cataloguing-in-Publication Data
A catalogue record for this book is available from the British Library

Additional hard and PDF copies can be obtained from orders@intechopen.com

Update on Malacology
Edited by Sajal Ray and Soumalya Mukherjee
p. cm.
Print ISBN 978-1-83969-743-2
Online ISBN 978-1-83969-744-9
eBook (PDF) ISBN 978-1-83969-745-6

For EU product safety concerns:
IN TECH d.o.o., Prolaz Marije Krucifikse Kozulić 3, 51000 Rijeka, Croatia, info@intechopen.com or visit our website at intechopen.com.

Meet the editors

Sajal Ray received an MSc in Zoology and MPhil in Environmental Science from Calcutta University, India, and a Ph.D. from Jadavpur University, India. His thesis reported the immunotoxicity of pesticides in an economically important snail of India. As a recipient of the Fogarty Visiting Fellowship, Dr. Ray carried out his postdoctoral research in cardiac pathology at the National Institutes of Health, USA. His research interest is studying the immunological responses of molluscs, sponges, crabs, and earthworms exposed to pollutants. His team is engaged in understanding the evolutionary mechanism of immunity in phylogeny. He has presented his research at various conferences including the World Congress of Malacology, Washington DC. Dr. Ray is currently a Professor of Zoology at Calcutta University.

Soumalya Mukherjee obtained an MSc degree in Zoology from West Bengal State University, India. He was selected for the DST INSPIRE Fellowship from the Department of Science and Technology, Government of India. He received his Ph.D. from the University of Calcutta, India. Dr. Mukherjee has published several research articles in peer-reviewed scientific journals and books of international repute. He has presented his research findings at various international and national conferences and symposia. He has been actively engaged in research of immunotoxicological effects of environmental pollutants on terrestrial and aquatic invertebrates. He was previously an assistant professor of Zoology, the Directorate of Open and Distance Learning, University of Kalyani, India. Dr. Mukherjee is currently Assistant Professor of Zoology, Brahmananda Keshab Chandra College, West Bengal State University, India.

Contents

Preface

Over the last few decades, the science of zoology has been witnessing a paradigm shift in research on malacology, which is the study of molluscs. Scientists of diverse disciplines are showing their research interest in molecular biology, cognition physiology, paleoecology, and other areas of malacological science. Their works provide us with a better understanding of the evolution, ethology, survival strategy, and molecular physiology of Mollusca, the second-largest phylum on the planet. In general, molluscs exhibit a marked variation in terms of body plan, adaptation, ecology, and physiological organization.

This book is a compilation of high-impact research articles on the frontier areas of molluscan biology, physiology, aquaculture, and paleoecology. The edited volume is representative of the current trends in global malacological research. It also highlights the basic and applied significance of molluscs inhabiting diverse habitats.

Chapters discuss the role and implications of different species of molluscs on nutrition, ecophysiology, farming, and paleoecology. As animals inhabiting terrestrial, freshwater, estuarine, and marine environments, molluscs have ecological, economical, evolutionary, biotechnological, cell biological, and paleontological significance. constitutes. They exhibit variation in morphology, body colour, food preference, and physiology. This book focuses on the various aspects of malacological research pursued in different parts of the world. Topics covered include effects of dietary intake of shellfish in humans, beneficial effects of herbal compounds on the cognitive ability of molluscs, seasonal variation of molluscs acting as intermediate hosts of human parasites, current understanding of freshwater pearl culture, and the role of environmental parameters on the infectivity of freshwater snails and their paleoecological aspects.

This book is a collection of informative articles written by scientists who are experts in the field. It is a rich source of information for students and researchers working in basic and applied malacology.

Sajal Ray
Aquatic Toxicology Laboratory,
Department of Zoology,
University of Calcutta,
West Bengal, India

Soumalya Mukherjee
Department of Zoology,
Brahmananda Keshab Chandra College,
West Bengal, India

Chapter 1

The Effects of Shellfish Consumption Frequency for Human Health

Latife Ceyda Irkin

Abstract

Depending on the world population, the importance of water resources and the consumption of aquatic organisms as a food source are increasing day by day. The presence of eicosapentaenoic acid (EPA) and docosahexaenoic acid (DHA), which are involved in critically important biochemical and physiological processes in the body, emphasizes the importance of seafood consumption. Shellfish are low in calories but rich in protein and omega-3 fatty acids. They also contain high amounts of many micronutrients, including iron, zinc, magnesium and B12. Consuming shellfish regularly can boost immunity, aid weight loss, and support brain and heart health. However, shellfish is one of the common food allergens, and some species may contain contaminants and heavy metals. Aquatic products poisoning occurs with the consumption of unhealthy seafood or fish containing toxins. Symptoms cause severe and fatal poisoning in consumers, depending on the presence and concentration of the toxin. To prevent food poisoning, information on the growing conditions of the species should be provided and regularly inspected for toxins (heavy metal poisoning and allergic reactions).

Keywords: Shellfish, health, diet, consumption

1. Introduction

Shellfish include shrimp, crayfish, crab, lobster, oysters, scallops, and mussels. Shellfish have been prepared in different ways and consumed for many years. Shellfish are animals that live in water and have a shell or shell-like exterior. They can be divided into two groups as crustaceans and mollusks. Crustaceans include shrimp, crayfish, crab, and lobster; oysters, scallops and mussels are examples of mollusks. Most shellfish live in saltwater, but the name also refers to species found in freshwater [1].

Shellfish are an important component of global seafood production. Shellfish contain a variety of vitamins and minerals, especially digestible proteins, and essential amino acids. It is among the foods that provide health benefits to consumers [2]. Although shellfish are generally a safe food source for consumption, they sometimes pose health risks due to their exposure to various habitats, the filtering of water by organisms such as oysters and mussels, and unhealthy agricultural practices.

Environmental hazards include factors that cause negative effects on living things and even cause death, especially pathogenic organisms and biotoxins. Appropriate preventive measures should be taken at the various stages of harvest, processing, storage, distribution, and consumption. For this reason, control measures are very important to protect the nutritional value and health benefits of shell products and consumer safety regarding the products [3, 4].

2. Some aquatic species with an economic aspect

Aquaculture can be obtained almost anywhere there is water. Especially in regions where temperate climate is dominant, the diversity of these products is increasing. While only fish and its derivatives are consumed in some places, less preferred foods such as octopus and even plants grown in the sea are consumed in some cuisines of the world.

Aquaculture means the products obtained from the species such as fish, mollusks, crustaceans, mammals, reptiles, sponges, and aquatic plants produced naturally or artificially in seas, inland waters and artificial pools, dams, ponds, fisheries, and fishing facilities. Some of the economical shellfish species found in the seas are explained below [5].

Gastropods; Sea snail (*Rapana thomasiana,* Gross 1861) (**Figure 1**) [6].

Cephalopods; Octopus (Octopus vulgaris, Linnaeus 1758, **Figure 2**), The European Squid (Loligo vulgaris, Lamarck 1798), The Common Cuttlefish (Sepia officinalis, Linnaeus 1758) [7].

Bivalves; Mussel (Mytilus galloprovincialis), Oyster (Ostrea edulis), Akivades (Tapes decussatus), Kidonia (Venus verrucosa), Sand mussel (Venus gallina) [8, 9].

Crustaceans; Shrimp (*Penaeus keathurus*), Insect (*Palinurus vulgaris*), Crayfish (*Astacus leptodactylus*), Spiny crab (*Maia squinada*), Blue crab (*Callinectes sapitus*), Lobster (*Homarus gammarus*) [10–13].

2.1 *Mytilus galloprovincialis*

The black mussel, whose scientific name is *Mytilus galloprovincialis* (Lamarck, 1819), is also known as the Mediterranean Mussel. The Black mussel is in the Bivalvia (bivalves) class of the Mollusca (**Figure 3**).

Mussels are creatures consisting of bivalves, interlocked with each other with a very strong musculature, triangular in front, ovoid in the back and bilateral

Figure 1.
Rapana thomasiana, *Gross 1861 [6].*

Figure 2.
Octopus vulgaris, *Linnaeus 1758 [7]*.

Figure 3.
Mytilus galloprovincialis, *Lamarck, 1819 [8]*.

symmetry. The shell consists of anterior margin, posterior margin, ventral margin, and dorsal margin. The leading margin is very brief and the shells interlock here. The outside of the shell is in diverse trace of purple-black and brown, and the inside is of pearl shine. There are growth lines on the shells that draw small elliptical circles starting from the junction of the shells. Although the common height of this species is 5–8 cm, it can reach a maximum of 10–11 cm. The temperature and salinity of the region have an important effect on the development of mussels, and the most delicious period of mussel meat is from autumn to the beginning of spring. Since bivalve mollusks feed by filtering water, they accumulate microbiological, chemical, or natural toxins in polluted aquatic environments, therefore, they must be given for consumption in a controlled manner. In aquatic environments with increased pollution, aquaculture rather than hunting is the most reliable way to enable controlled production [14–16].

Black mussel is a popular seafood that is consumed with admiration all over the world, especially in European and Pacific countries. These creatures, which live in the coastal regions of the marine ecosystem, can be obtained from nature by hunting or they can be produced through culture. Mussels, which have as valuable and quality protein as fish meat, can be consumed in various ways such as stuffed mussels, pan and brine. Mussels, which have a short shelf life, should be prepared, and consumed as soon as possible or stored by paying attention to storage conditions [16].

2.2 *Penaeus keathurus*

Shrimp are ten-legged arthropods that live in fresh and salt water. Freshwater shrimps are mostly common in tropical regions. Their size is varying cylindrical body between 1 and 30 cm. Its body is covered with an armor made of calcium carbonate. These creatures, whose bodies are jointed, swim backwards by waving their wide, fin-like tails. It has five pairs of legs and at least two of them have claws. The antennae, which are two pairs, are very long and bifurcated. At least one of these forks bends back, allowing the shrimp to retract into the crevices and signal danger from behind. The most obvious reaction of the shrimp in the face of danger is to try to protect itself with a sudden twist. Their body color can change to suit their environment. There are large-clawed predators that feed on small fish, as well as scavengers that feed on food particles in the sand. In some species, there are brush-like bristles on their claws to easily collect food particles (**Figure 4**) [17].

Some types are less tasty than others and some types are not consumed. It is farmed in countries such as America, Japan, Thailand, and Taiwan. Consumption in the USA is higher than in other countries. Since shrimp species found in cold waters grow slowly, their meat is more delicious. Depending on the species, their flesh can be firm and transparent, pink, yellow, gray, brown and red. The flesh color, which is transparent when cooked, takes a dull and pinkish color. One of the commercially important species is the deep-water shrimp (*Pandalus borealis*). The other is the large black shrimp (*Penaeus monodon*). Since shrimp species spoil easily after hunting and melanosis (black spots) form on their meat, they should be cooled immediately, and their heads should be cut off. The most expensive of the shrimp species are the largest ones. If fresh shrimp are to be bought, their bodies should be firm, not sticky, and soft, their bodies should not be separated from their shells and their heads should not be surrounded by black spots. It can be stored fresh for about 2 days in the refrigerator and frozen for 1 month [18].

2.3 *Loligo vulgaris*

Squid, like other cephalopods, has a prominent head, bilaterally symmetrical structure, mantle, and arms. The substance called melanin in it is the same pigment that tans human skin. Its eyes are in the middle of its head and body. It has 8 arms surrounding the mouth with a sharp, parrot-like beak, and two distinctly longer tentacles. Most of the squid is formed by a thick muscle cover called the mantle, which protects the internal organs and allows the squid to move through the water by spraying strongly with water. It compresses the water in the mantle and sprays the water quickly from the section also called siphon, allowing it to swim backwards. Thanks to its funnel-shaped structure that creates a jet effect, squid can accelerate in water more than 3 times the speed of Olympic swimmers (**Figure 5**) [19].

Figure 4.
Venus verrucosa, *Linnaeus, 1758 [9]*.

Figure 5.
Penaeus keathurus, *Forskål, 1775 [10].*

Squid, which is the lowest calorie seafood product, contains protein, very little fat, phosphorus, magnesium, and calcium. It also contains vitamins B2, B3 and B12. To clean the squid, the internal organs must be removed by pulling from the tail. By holding the tip of the squid, it is necessary to quickly pull and clean the outer skin. After cleaning the outer skin and internal organs, the transparent bone in the middle of the squid should also be removed and the squid should be washed well. The cleaned squid is offered for sale in the form of rings and frozen. Fresh squid can be stored in the refrigerator for 1–2 days. Frozen squid can be stored in the freezer for 1–2 months [20].

2.4 *Ostrea edulis*

Ostrea edulis (Linnaeus, 1758) is a bivalve mollusk with very tasty meat and cultivated. It consists of two circular shaped shells and these shells are connected to each other by a structure called ligament. It lives in offshore sandy, pebbly, or rocky areas in all our seas. It is not found in brackish waters. It feeds on plankton and suspends organic matter. They are oysters evaluated as fresh. The peel is quite light, thick, and oval. Yellowish-brown, right bark is flat, straight, and covered with inconspicuous radial folds. The left shell is in the form of a convex cube and its edges are serrated. The shell surface is irregularly indented. Its maximum length is 12 cm. It has a characteristic shellfish odor. There are no irritating odors (ammonia smell, etc.) (**Figure 6**) [21].

The shells do not open with manual intervention, but they can be opened by cutting the ligament with the help of a cutter. When it first reaches sexual maturity, the gonad normally develops like a male and gives off sperm. After the gonad releases the sperm, it passes into the female stage and produces eggs instead of sperm. This

Figure 6.
Callinectes sapidus, *Rathbun, 1896 [11, 12].*

continues a regular basis throughout his life. The formation of pearls is completed because of the combination of sand and similar materials in the seas, where many kinds of creatures can live, by entering the oyster shells and combining with the mother-of-pearl secretions. Fresh products can be stored in cold stores between 0°C and + 4°C between pieces of ice. Frozen products that have been frozen at −40°C and will wait for a long time should be stored at a temperature of −22°C to −18°C in the center. The product should not be kept together with substances that emit a foreign odor or that will pollute it. Meat texture should be fresh, firm, and unique, natural color. It is served as fresh raw or cooked with various sauces. It is recommended to boil it with its peel during cooking. It is offered for sale as cooked canned or smoked oysters [22].

2.5 *Astacus leptodactylus*

Although the Eastern European crayfish (*Astacus leptodactylus*) was considered an important commercial product (a luxury food item) in the world after the 1830s, it was only used in World War II in Turkey. After the World War II, it became one of the important export products among aquaculture products. While the total crayfish production was 500 tons in 1979, this value was 6500 tons in 1982, and after 1986, it gradually decreased due to the occurrence of crayfish plague (*Aphanomyces astaci*), overfishing and environmental pollution. Today, total crayfish production (1894 tons, 2002 data) is around 15% of the 1980s. About 600 crayfish species are found naturally in other continents except Africa and Antarctica. In addition, crayfish migrate from their original environment to other environments, intensively, naturally (with migration or currents); They were transported by chance (by ships' ballast waters, channels, being used in traps for catching fish, escaping from the environments where they were kept under control, being carried by predators or people unknowingly) or consciously by people (keeping them as a hobby in aquariums, production and aquaculture, control of aquatic plants). The most important and common factor in the transportation of crayfish from one environment to another is the desire of entrepreneurs to earn economic income from these creatures (**Figure** 7) [23].

The trunk is divided into two parts, the thorax, and the abdomen. The chest is covered with a hard and prominent shell. The skeleton-shell is outside. The shell is immobile and hard except for the articular parts. The joints are soft and thin, and their structure is also different. There are 4 pairs (8 pieces) of feet for walking in the chest part. The abdomen consists of 6 segments. Freshwater crayfish are smaller

Figure 7.
Astacus leptodactylus, *Eschscholtz, 1823 [13].*

and red in color. Those that live in salt water are also larger and lighter in color. They are voracious and aggressive species. It is found in rivers, lakes, streams, and ponds. In many countries, the consumption of crayfish with pleasure and the increase in its economic value day by day has accelerated the production of this product under cultural conditions. It is cultivated on farms in Europe, especially in South America. There are 300 different types available. Some species are distinguished by the red or white color of their claws. Their shells are red, brown, and purple in color. It has lean and delicious pink, white flesh. Crayfish normally have a heavy, wobbly gait. Walking is forward through the legs. However, their swimming is backwards. Crayfish generally like flowing and abundant calcareous waters. It enables the development of limestone crusts. They are rarely found in acidic waters. It can be sold live, cooked, frozen, or canned. If cooked, those with firm skins and complete claws should be chosen. It must be cleaned before cooking. It contains very little meat. Live crayfish can be stored wrapped in a damp cloth for 12 hours in the refrigerator, 1–2 days in the refrigerator when cooked, and 1–2 months frozen [24].

3. The importance of nutrition with shellfish

Shellfish are rich in low-calorie lean protein, essential oils, and micronutrients. Most of the fats found in shellfish are in the form of omega-3 fatty acids, which have benefits for improving brain and heart health. Shellfish are rich in iron, zinc, magnesium, and vitamin B12, all of which play important roles in our bodies. For example, 85 g oysters have almost 100% of the daily value for zinc. Shellfish are most nutritious when steamed. Fried shellfish may contain ingredients such as additional calories, refined carbohydrates, added salt. With its impressive nutritional content, shellfish has low calories. This makes them excellent foods to eat while trying to lose weight. Protein-rich foods can help you lose or maintain weight by preventing you from consuming calories. Due to its omega-3 fatty acid content, it can lead to a greater feeling of satiety than fish and may help to lose weight faster than other high-protein foods. A study of overweight adults found that those who consumed more omega-3 fatty acids on a calorie-restricted diet felt significantly fuller after meals than those who consumed less omega-3 on the same diet [25]. In order to obtain the maximum benefit from shellfish as food, it is extremely important that the environmental conditions in which the consumed products are supplied are healthy, as well as the depuration of the products by the companies and the storage conditions of the packaged products until the expiry date without breaking the cold chain.

Shellfish have microelements that can improve the health of heart, including omega-3 fatty acid and vitamin B12. Studies have shown that getting omega-3 fatty acid from fish and shellfish is linked to a lower risk of heart disease. This is probably because omega-3 have anti-inflammatory effects. In a study of 18,244 healthy men in China, those who ate more than 200 grams of omega-3-rich shellfish per week were 59% less likely to die from a heart attack than those who ate less than 50 gram per week. Inadequate B12 intake has also been linked to high blood levels of homocysteine, a protein that may increase your risk of heart disease. Therefore, foods rich in vitamin B12 may protect against heart disease. Studies have identified insufficient B12 and omega-3 levels as risk factors for problems with brain development in children and healthy brain function in adults. Some research also suggests that vitamin B12 and omega-3 fatty acid may improve each other's activities to improve brain health. In a study of 168 adults with mild mental disorders it was found that B vitamins slowed the progression of brain problems in those with lower levels of omega-3 fatty acids compared to those with low levels. Shellfish contain zinc, which

Type	Calories	Protein	Fat
Shrimp	72	17 grams	0.43 grams
Crayfish	65	14 grams	0.81 grams
Crab	74	15 grams	0.92 grams
Lobster	64	14 grams	0.64 grams
Clams	73	12 grams	0.82 grams
Scallops	59	10 grams	0.42 grams
Oysters	69	8 grams	2 grams
Mussels	73	10 grams	1.9 grams

Table 1.
Nutrition facts of 85-gram of different types of shellfish (https://fdc.nal.usda.gov/ndb/).

strengthens the immune system. This mineral is necessary to strengthen the cells that make up the immune defense of our body. It also acts as an antioxidant, protecting against damage caused by inflammation. A study of 62 healthy adults over the age of 90 showed that zinc deficiency reduced the activity of certain immune cells. Shellfish are full of protein and healthy fats that can aid weight loss. It is also rich in omega-3 fatty acids, vitamin B12 and zinc, which support a healthy brain, heart, and immune system (**Table 1**) [26].

4. Negative effects of excessive consumption of shellfish

Although shellfish are highly nutritious, they have some disadvantages when consumed excessively. Shellfish have the potential to accumulate heavy metals such as mercury or cadmium from the environment. The accumulation of these compounds in our bodies can lead to organ damage and other health problems. One study showed that shellfish in some regions may contain cadmium at twice the recommended daily dose. The Food and Drug Administration recommends that adults eat 85–140 grams of low-mercury fish twice a week. If the number of shellfish we eat in a week is equal to or less than this, there is no danger in terms of heavy metals. Consuming shellfish obtained from waters where polluting factors prevail causes many food-borne diseases. Mollusks such as oysters and mussels accounted for more than 45% of seafood-related foodborne illness cases in the United States America (USA) from 1973 to 2006. Food poisoning from shellfish can be caused by bacteria, viruses, or parasites in the environment. Pathogens thrive in raw shellfish that are not properly refrigerated. Therefore, obtaining shellfish from clean waters, storing, and cooking them properly is the most effective way to prevent foodborne illness. Pregnant and nursing mothers, older adults, and people with compromised immune systems should avoid raw or improperly prepared shellfish [27, 28].

Shellfish is one of the top eight food allergens in the USA. Shellfish allergy typically develops in adulthood but can also occur in childhood. Symptoms of an allergic reaction to shellfish include vomiting and diarrhea, stomach pain and cramps, swelling of the throat, tongue or lips, hives, shortness of breath. In some cases, people with a shellfish allergy may experience a life-threatening anaphylactic shock that requires immediate treatment. Shellfish can accumulate different levels of heavy metals that can build up in your body and cause health problems. Diagnosing allergies to shellfish can be difficult and complex. Symptoms can vary from person to person, depending on metabolism, and may not always give the same reaction. People with shellfish allergies may not need to eat only these products to develop a reaction. A reaction may also occur on contact with cooked shellfish. Allergic reactions to these creatures can also affect the skin, respiratory, digestive, and circulatory systems. Although these allergies are not very common in adults, they can occur at any age. In the case of this type of food allergy situation, it is important to consult an allergist who can determine what tests need to be done after diagnosis, decide whether it is an allergy, and advise patients on how to manage exposure and symptoms [29].

Allergies to shellfish occur when the immune system opposes the proteins found in these animals. When these proteins enter the body of individuals with allergies, the immune system overreacts and tries hard to fight against the antigens it sees as foreign. Some of these reactions occur due to the release of histamine, which causes allergy symptoms. Therefore, antihistamines can give effective results in such an allergic reaction. For people whose shellfish allergy is confirmed by further testing, the only treatment is to avoid these creatures altogether. In the event of exposure or a serious reaction, every second counts. Detecting the reaction early and administering epinephrine quickly can prevent worsening of the condition and possible death [30, 31].

History is very important for diagnosis in patients with suspected Shellfish allergy. The patient's history is necessary both for determining the severity of the disease and for planning diagnostic procedures. Therefore, before going to the doctor, review your complaints in detail. In general, those with suspected allergies,

- Skin prick tests,

- Determination of fish-specific IgE in serum,

- If necessary, loading tests are performed with foods that are thought to be responsible. In recent years, component-based diagnosis method has also been applied to make a more accurate diagnosis and obtain information about the course [32].

5. Conclusion

Shellfish, which can be divided into crustaceans and mollusks, are full of lean protein, healthy fats, and many beneficial microelements. They can aid weight loss, boost immunity, and improve brain and heart health. However, shellfish may contain heavy metals. May cause foodborne illness and allergic reactions. However, shellfish are delicious foods that support a balanced diet. Shellfish are among healthy foods when consumed in moderation.

Shellfish is one of allergens covered by the labeling of the FALCPA (Food Allergen Labeling and Consumer Protection Act). This labeling must appear on packaged food products containing shellfish sold in the USA. The presence of shellfish in the product should be indicated on the packaging. Shellfish is rarely hidden in consumption. Shellfish may be found in fish stocks, flavoring, surimi and sushi. Anyone with a food allergy should read ingredient labels and take precautions. Doctors can point you to helpful resources that can help you plan your meals, such as patient support groups and registered dietitians.

Conflict of interest

No conflict of interest.

Author details

Latife Ceyda Irkin
Applied Sciences Faculty, Fisheries Technology Department, Çanakkale Onsekiz Mart University, Çanakkale, Turkey

*Address all correspondence to: latifeirkin@gmail.com

References

[1] Rodger C, Lees D, Hudson S. Classification and monitoring of shellfish harvesting areas in England and Wales; 1992.

[2] U.S. Seafood Consumption Declines Slightly in 2009. http://www.noaanews.noaa.gov/stories2010/20100909_consumption.html]

[3] Sicherer SH, Munoz-Furlong A, Sampson HA. Prevalence of seafood allergy in the United States determined by a random telephone survey. J Allergy Clin Immunol. 2004;114(1):159-165.

[4] Chiang WC, Kidon MI, Liew WK, Goh A, Tang JP, Chay OM. The changing face of food hypersensitivity in an Asian community. Clin Exp Allergy. 2007;37(7):1055-1061.

[5] Per C, Maryam R, Amadou A et al. Sustainable fisheries and aquaculture for food security and nutrition. 2014.

[6] https://commons.wikimedia.org/wiki/File:R apana_Black_Sea_2008_G1.jpg, Access: 10.09.21

[7] https://www.biolib.cz/en/image/id363305/, Access: 10.09.21

[8] Filipa P. Ship transport of marine invasive species and its stress resistance. Master thesis. 2014.

[9] https://commons.wikimedia.org/wiki/File:Venus_verrucosa.jpg, Access: 10.09.21

[10] José G, Santana J.I. The family Penaeidae from the Canary Islands (Northeastern Atlantic), with first record of *Penaeus kerathurus*. Boletim do Museu de História Natural do Funchal. 2014;64: 29-34

[11] http://europeantrackingnetwork.org/europe an-squid-loligo-vulgaris-3, Access: 16.09.21.

[12] https://www.istockphoto.com/tr/foto%C4%9Fraf/blue-crab-gm117002934-6483213, Access: 10.09.21

[13] http://dogalhayat.org/property/astacus-leptodactylus, Access: 10.09.21

[14] Eleonora V, Herbert WJ. Yield and Post-Yield Behavior of Mussel Byssal Thread: A Self-Healing Biomolecular Material. Biomacromolecules. 2001;2(3):906-911.

[15] Northern Economics, Inc. The Economic Impact of Shellfish Aquaculture in Washington, Oregon and California. Prepared for Pacific Shellfish Institute. Retrieved November 30, 2018.

[16] Buehler QZ, Markus J. Impact tolerance in mussel thread networks by heterogeneous material distribution. Nature Communications. 2013;4:2187.

[17] Manaşırlı M, Özyurt C, Kıyağa V, Avşar D. The Growth Parameters of the Mediterranean Shrimp (*Penaeus kerathurus* Forskal, 1775) in the Iskenderun Bay. Ecological Life Sciences. 2018;13(1):15-26.

[18] Rudloe J, Rudloe A. Shrimp: The Endless Quest for Pink Gold FT Press. 2009. ISBN 9780137009725.

[19] Tricarico E, Amodio P, Ponte G, Fiorito G. Cognition and recognition in the cephalopod mollusk *Octopus vulgaris*: coordinating interaction with environment and conspecifics. In Witzany, G. (ed.). Biocommunication of Animals. Springer. 2014; 337-349

[20] Faculty of Science - University of Copenhagen. Cephalopods could become an important food source in the global community. ScienceDaily. Retrieved August 9, 2021 from www.sciencedaily.com/releases/2018/10/181029130954.htm

[21] Brian M. Bivalve. Encyclopedia Britannica. https://www.britannica.com/animal/bivalve. Accessed 9 August 2021.

[22] Carlton JT. Molluscan invasions in marine and estuarine communities. Malacologia. 1999;41(2):439-454.

[23] Balık S, Ustaoglu M, Sarı H, Berber S. Determination of traits some growth and morphometric of crayfish (*Astacus leptodactylus* Eschscholtz, 1823) at Demirköprü Dam Lake (Manisa). Su Ürünleri Dergisi. 2005;22. 10.12714/egejfas.2005.22.1.5000156891.

[24] *Astacus leptodactylus*-Turkish Crayfish. UK non-native organism risk assessment scheme version 3.3. DEFRA. Archived from the original on 25 February 2019.

[25] Seafood Nutrition Partnership http://www.seafoodnutrition.org/

[26] Hosomi R, Yoshida M, Fukunaga K. Seafood consumption and components for health. Global Journal of Health Science. 2012;4(3):72-86. https://doi.org/10.5539/gjhs.v4n3p72

[27] https://fdc.nal.usda.gov/ndb/

[28] James KJ, Carey B, O'Halloran J, van Pelt FN, Skrabáková Z. Shellfish toxicity: human health implications of marine algal toxins. Epidemiol Infect. 2010;138(7):927-940.

[29] Sicherer SH, Munoz-Furlong A, Sampson HA: Prevalence of seafood allergy in the United States determined by a random telephone survey. J Allergy Clin Immunol. 2004;114 (1):159-165.

[30] Wai CYY, Leung NYH, Chu KH, Leung PSC, Leung ASY, Wong GWK, Leung TF. Overcoming Shellfish Allergy: How Far Have We Come? Int J Mol Sci. 2020;23;21(6):2234.

[31] Khora SS. Seafood-Associated Shellfish Allergy: A Comprehensive Review. Immunol Invest. 2016;45(6):504-530.

[32] NIAID-Sponsored Expert Panel et al. Guidelines for the diagnosis and management of food allergy in the United States: report of the NIAID-sponsored expert panel. The Journal of allergy and clinical immunology. 2010;126(6):S1-58. doi:10.1016/j.jaci.2010.10.007.

Chapter 2

Green Tea-Derived Catechins Have Beneficial Effects on Cognition in the Pond Snail

Yoshimasa Komatsuzaki, Ayaka Itoh and Minoru Saito

Abstract

Green tea has been used as a medicine in East Asia for thousands of years. Plant-derived compounds called flavanols, which are included in green tea, may have potentials to help maintain healthy brain function. In this chapter, we review the effects of flavanols, e.g. epicatechin (EpiC), on cognitive ability in the pond snail, *Lymnaea stagnalis*. In this decade, the Lukowiak's group has tested the effects of EpiC on cognition ability in *Lymnaea*. In a *Lymnaea* model system, they showed that EpiC and EpiC-containing foods have a rapid and activity-dependent effect enhancing the formation of long-term memory (LTM) following operant conditioning of aerial respiratory behavior. In the last part of this chapter, we also introduce our study for the effects of EpiC on LTM formation in another model system in *Lymnaea*. This study showed that EpiC increases the persistence of LTM formed by classical conditioning of feeding behavior, and suggested that EpiC alters some electrophysiological properties of a neuron in the feeding system.

Keywords: Green tea-derived catechins, Epicatechin, Operant conditioning, Classical conditioning, Learning and memory, Long-term memory, *Lymnaea*

1. Introduction

Green tea is one of the most popular beverages in the world. It is made from *C. sinensis* leaves and include many kinds of phytochemicals. Compounds called flavanols, which are included in green tea, are candidates for the active ingredient that has been used as a medicine in East Asia for thousands of years. An extract, called Sinecatechins, of green tea leaves is also used as botanical drug approved by the FDA in USA [1]. Flavanols (catechins) belonging to the group of polyphenols [2] are contained in green tea (mg/100 g): 26.05 (−)-epigallocatechin-3-gallate (EGCG), 7.57 (−)-epicatechin-3-gallate (ECG), 16.02 (−)-epigallocatechin (EGC) and 6.16 (−)-epicatechin (EpiC) [3]. Recently, it is reported that flavanols may have potentials to help maintain healthy brain function in both vertebrates and invertebrates [4–8].

Gastropod mollusks such as *Aplysia*, *Limax*, *Hermissenda* [9–14] and *Lymnaea* [15–19] are excellent model animals for understanding the causal neuronal mechanisms of learning and memory. In this decade, the Lukowiak's group in University of Calgary has tested the effects of EpiC on cognition in *Lymnaea stagnalis.* In a *Lymnaea* model system, they showed that EpiC and EpiC-containing foods have a

 IntechOpen

rapid and activity-dependent effect enhancing the formation of long-term memory following operant conditioning of aerial respiratory behavior.

Lymnaea is an aquatic pulmonate snail and can breathe either with its lung or skin. It approaches water surface and gets air into the lung through opening its pneumostome (**Figure 1A**). The Lukowiak's group employed a protocol of operant conditioning of aerial respiration to investigate the cognitive function in *Lymnaea*. In a hypoxic environment in which the frequency of aerial respiration in *Lymnaea* increases, applying repeated tactile stimulus to pneumostome as a negative reinforcement (training session; TS) reduces the number of attempted pneumostome openings in *Lymnaea*, and then the behavioral change persists for 3 hours or longer.

A single 30-min training session (0.5 h-TS) results in intermediate-term memory (ITM) that persists for up to 3 hours, whereas two 30-min training sessions with a 1 hour rest interval (2 h-TS) results in long-term memory (LTM) that persists for 24 hours [21]. ITM depends on the translation of existing mRNA transcripts but does not require mRNA transcription. LTM requires both the translation of mRNA and the formation of new mRNA transcripts [22]. Thus, a 2 h-TS could drive the process of mRNA transcription in addition to translation of mRNA.

To drive aerial respiration in *Lymnaea*, a 3-neuron central pattern generator (CPG) was shown to be both necessary and sufficient (**Figure 1B**) [23, 24]. Subsequently, it was shown that one of the three CPG neurons, Right Pedal Dorsal 1 (RPeD1), is a necessary site for LTM formation, extinction and reconsolidation of the memory [18, 22, 25]. It is also possible to utilize a semi-intact preparation where aerial respiratory behavior and neuronal activity can be simultaneously studied [26, 27].

Lymnaea can be classically, as well as operantly, conditioned and LTM can be formed by the following learning procedures [28, 29]. Conditioned taste aversion (CTA), which is a classical conditioning, is based on pairing sucrose as a conditioned stimulus (CS) with an aversive chemical unconditioned stimulus (UCS) such as KCl, which inhibits feeding and evokes a withdrawal response. After ten trials, the feeding response of trained snails to sucrose became significantly weaker than that of control snails, and this associative memory lasted for more than 2 weeks [30].

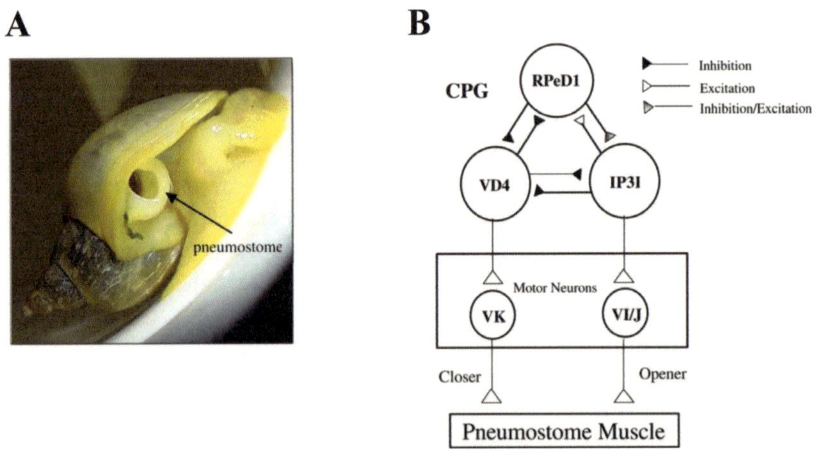

Figure 1.
Pneumostome in Lymnaea *and the neural circuit including in aerial respiratory behavior. (A)* Lymnaea *with the opened pneumostome. (B) Schematic drawing of the central pattern generator (CPG) to drive aerial respiration. Depolarization of Right Pedal Dorsal 1 (RPeD1) activates input3-interneuron (IP3I) via a biphasic effect (inhibition followed by excitation). Subsequently, IP3I excites both RPeD1 and a group of motor neurons (VI/J cells) involved in pneumostome opening. IP3I also inhibits visceral dorsal 4 interneuron (VD4), which is involved in pneumostome closing. The combined inhibitory input from both RPeD1 and IP3I causes burst firing of VD4. These figures are reproduced from [20] with permission.*

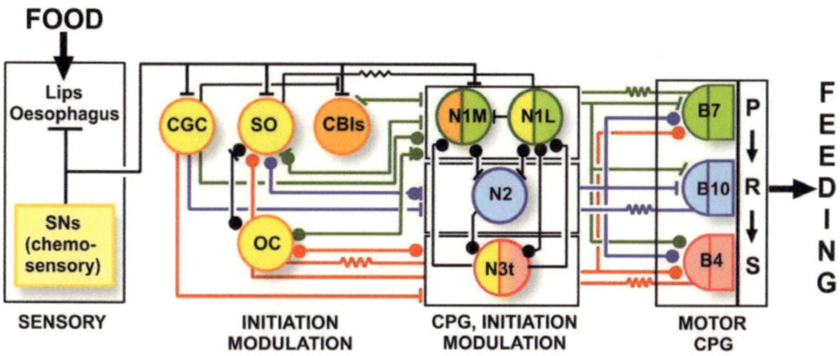

Figure 2.
Neuronal circuit corresponding to the feeding behavior. Each neuron is indicated by an abbreviation (see [31]). Modulatory function is indicated by yellow and initiating function by orange. Central pattern generator (CPG) interneurons and motoneurons active during the three phases of the feeding rhythm are indicated by green (P = protraction), blue (R = rasp) and red (S = swallow). Neurons labeled with two colors have two functions. Dots indicate inhibitory chemical synapses, bars excitatory chemical synapses and resistor symbols electrotonic (electrical) synapses. This figure is reproduced from [31] with permission.

Individual neurons in *Lymnaea* can be identified as the neuronal circuit corresponding to the feeding behavior (**Figure 2**). An identified spontaneously active pair of neurons, the cerebral giant cells (CGCs), has been shown to both modulate the neuronal network underling the feeding behavior and be necessary for LTM and its retrieval following CTA training [31, 32]. The most significant CGC synaptic connections are with the neuron 1 medial (N1M) cell, an interneuron in the CPG that co-ordinates rhythmic feeding movements [33–36].

In the last part of this chapter, we introduce our study for the effects of EpiC on LTM formation in the feeding system in *Lymnaea*.

2. Enhancing effects of epicatechin on memory formation by operant conditioning of aerial respiratory behavior

To date, the most detailed information on effects of EpiC in *Lymnaea* has been obtained from experiments using operant conditioning of aerial respiratory behavior. In this paradigm, snails are subjected to a protocol that the pneumostome receives a weak tactile stimulus whenever the snail attempts to open the pneumostome in a hypoxic environment. The number of attempted pneumostome openings is recorded for each snail for 30 minutes. To determine whether memory is formed following the training session (TS), an identical procedure is performed 24 hours later, which is called a memory test (MT). The number of attempted pneumostome openings in the MT is compared with that in the TS, and long-term memory (LTM) is evaluated if the number of attempted openings in the MT is significantly lower than that in the TS [25, 37].

When snails were given a 0.5 h-TS, which do not usually form LTM lasting 24 hours or more, in the presence of 15 mg/L EpiC, the memory persisted until 24 hours after that training. Thus, EpiC can promote LTM formation by driving the process of mRNA transcription in addition to mRNA translation. EpiC also enhances LTM formation. When snails were operantly conditioned in EpiC-containing pond water (15 mg/L) by a 2 h-TS (TS1, TS2; **Figure 3A**), which typically results in memory lasting only 24 hours in pond water without EpiC (control group, TS1 versus MT, n = 12, no significance; **Figure 3B**), they formed LTM lasting at least 72 hours (EpiC group, TS1 versus MT, n = 12, $P < 0.01$; **Figure 3C**) [4, 38]. Moreover, following a 2 h-TS in EpiC-containing pond water, snails were received the MT in standard pond water (i.e. no EpiC) at 96 hours, 1 week and 2 weeks

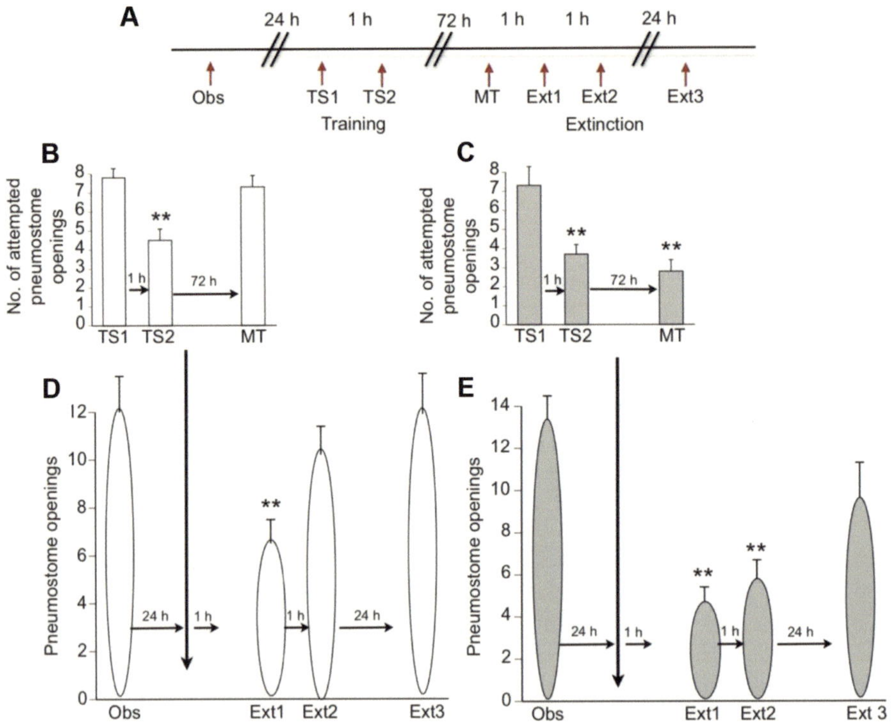

Figure 3.
*Epicatechin (EpiC) enhances LTM formation and diminish the rate of extinction following operant conditioning. (A) A timeline of the experiment is shown. (B–E) White and gray bars show snails trained without EpiC or with EpiC, respectively. Snails were either operantly conditioned in pond water (B) or EpiC-containing pond water (C). (D, E) In an observation (Obs) and extinction sessions (Ext1–3), the number of pneumostome openings was calculated for each snail without tactile stimulation in hypoxic water for 30 minutes. (B, C) n = 12, **P < 0.01 compared with TS1. (D, E) n = 12, **P < 0.01 compared with Obs. These figures are produced from [4] with permission.*

after the 2 h-TS [38]. The snails maintained in a very long-term memory lasting for 2 weeks or longer. Memory formed in the presence of EpiC is resistant to forgetting, which is not dependent on EpiC being present in the MT. In addition, snails received extinction sessions (Ext1–3). In extinction sessions, snails are allowed to freely perform aerial respiration in the same hypoxic environment as TS, and the learned association between the pneumostome opening and weak tactile stimulus can be extinguished. Snails trained in the presence of EpiC were also more resistant to extinction of memory (EpiC group, Obs (pre-training) versus Ext2, n = 12, $P < 0.01$; **Figure 3E**), including blocking “the original memory” by overwriting a newer memory [39, 40], than control snails (control group, Obs (pre-training) versus Ext2, n = 12, no significance; **Figure 3D**) [4].

Exposure to EpiC does not alter locomotor activity and spontaneous aerial respiratory behavior themselves in *Lymnaea* compared to naïve snails [4]. Thus, studying the effects of EpiC could exemplify specificities for drugs that directly interact with neuronal signaling pathways for LTM formation and persistent.

In mice, EpiC improves retention of spatial memory by enhancing angiogenesis [8]. Oral intake of EpiC via gavage increases the level of EpiC and its metabolites in rat plasma and brain [41], and then EpiC could influence brain functions. In *Lymnaea*, EpiC can easily absorb via skin into the body cavity, and subsequently may contact to the CNS by an open circulatory system.

Additionally, it is reported that EpiC also has a rapid and activity-depend effect on promoting LTM formation in *Lymnaea* [4, 5, 38]. Fernell et al. demonstrated that

EpiC must be present during operant conditioning or applied with 1 hour immediately after training in order for EpiC to cause LTM enhancement [42]. However, EpiC exposure at 1 hour after training did not result in enhanced memory formation. Thus, One hour after training is important for LTM formation. It is widely known that there is an important period for encoding and consolidation of the memory following learning [43]. In *Lymnaea*, the consolidation period persists for about 1 hour. LTM requires both regulation of gene activity and new protein synthesis, whereas ITM requires only new protein synthesis. Previously, it was shown in *Lymnaea* that applying a cold block (10 minutes in 4°C pond water) immediately after training blocks only LTM formation, but not block ITM formation [44, 45]. Thus, these results suggest that EpiC could alter the gene activity for LTM.

A specific mechanism by which EpiC enhances memory formation has not been clarified. However, there have been many studies in mammals suggesting various ways in which dietary flavonoids may exert such beneficial effects on the CNS [8]. EpiC is known as antioxidant, protecting neurons from injury caused by oxidative stress [46]. EpiC can be photo-inactivated by exposure to ultraviolet light (UV). Exposed to the sun for 6 hours, EpiC changes the molecular conformation by breaking the cyclic ether through a radical mechanism [47]. But no significant change was observed in the antioxidant activities of EpiC upon 6 h beta-UV radiation. Following photo-inactivation of EpiC, memory enhancement did not occur. Photo-inactivation of foods containing EpiC also blocked their ability to enhance LTM [7]. Thus, enhancing effect of EpiC on memory formation in *Lymnaea* is less likely to be caused by antioxidant properties of EpiC and may be directly due to affecting the signaling pathway required for memory formation in neurons. This does not mean that antioxidant properties do not act against oxidative damage and therefore protective effect may contribute to retain memory over long-term.

As mentioned above, LTM formation is dependent on altered gene activity and new protein synthesis [48]. It is well-known that the consolidation period following learning plays an important role for LTM formation [20, 49], in which the learning is encoded into memory. EpiC can cause an enhancement of memory formation if snails experience EpiC during training or immediately after training. However, EpiC exposure at 1 hour before training or 1 hour after training was not sufficient to cause memory enhancement [42]. In *Lymnaea*, it is thought that the consolidation period persists for about 1 hour. If the operant conditioning is performed in the presence of environment stressors (exposure to predator kairomones or KCl), it results in strengthening of memory formation. The effects via sensory input from the osphradium (a sensory organ) are dependent on a serotonergic signaling pathway [50]. However, the enhancing effects of EpiC do not require either the input from the osphradium or serotonergic signaling pathway.

3. Signaling pathway involved in epicatechin effect

EpiC effects on LTM formation may be due to its ability to drive an increase in intracellular kinase activity [51] in neurons such as RPeD1. It has also been shown that activation of CREB is necessary for LTM formation in *Lymnaea* [52] and EpiC increases CREB-regulated gene expression in neurons [53]. Further, there is increasing evidence implying that EpiC can drive rapid signaling intracellular as it increases phosphorylation of protein kinase B (Akt)/PI3K, PKC and Erk MAPK and induces cellular survival/proliferation in human hepatoma cells [54]. This is important since LTM following operant conditioning in *Lymnaea* requires activations of PKC and MAPK [55]. In addition, EpiC appears to be able to directly alter DNA methylation activity [56], which has been shown in *Lymnaea* to alter LTM formation [57]. EpiC has been shown in the mammalian brain to cross the blood brain barrier and directly affect CNS function possibly by enhancing 5HT function [58]. EpiC may also activate NOS and stimulate NO production [53]. It is known in the *Lymnaea* model system that 5HT and NO are involved in LTM formation [50, 59]. It remains to be elucidated whether EpiC brings about its enhancing effects on LTM formation via these molecules. EpiC effects on cognitive enhancement in mammalian preparations has been shown, but it is unclear whether the enhanced cognitive benefit is directly due to altering neuronal activity or through effects on blood flow to the brain as a result of increased angiogenesis [8].

It is reported that exposure to crayfish effluent (CE), which also enhances LTM formation and significantly decreases RPeD1 excitability [60], works a serotonergic pathway that can be blocked by mianserin, a serotonin receptor antagonist [50]. As previously shown, however, mianserin does not affect the LTM formation enhancement induced by EpiC [4]. In addition, once the osphradial nerve that connects the osphradium (a sensory organ) to the CNS is severed, CE no longer enhances LTM formation [50]. Thus, the osphradial nerve must be intact in order to cause enhancement of LTM formation by exposing CE. Whereas, EpiC enhanced LTM formation after severing the osphradial nerve [4]. Thus, it appears that EpiC acts via a different mechanism and a different pathway from those caused by the perception of CE.

McComb and collaborators demonstrated the memory formation by using *in vitro* semi-intact preparations [26]. After operant conditioning of intact snails, semi-intact preparations were dissected so that changes in the respiratory behavior (pneumostome openings) and underlying activity of the identified CPG neuron, RPeD1, could be monitored simultaneously.

Our group can perform "*in vitro*" operant conditioning in semi-intact preparations from naïve snails. In the training, we applied a gentle tactile stimulus to the pneumostome area whenever the snail began to open it. Following the training, the respiratory behavior decreased. After the training, naïve snails exposed to EpiC (15 mg/L) prior to recording exhibited significantly increased RPeD1 excitability compared with non-exposed snails. This experiment can help to understand how EpiC alters RPeD1 excitability to drive aerial respiratory behavior and leads to enhanced LTM formation.

These results provide the basis of future studies in *Lymnaea* to elucidate how EpiC enhances LTM formation of respiratory conditioning.

4. Effects of intaking of catechin-rich foods

Does exposure to food products containing EpiC during the training elicit similar effects seen for exposure to pure EpiC? Lukowiak et al. demonstrated interesting experiments whether foods containing substantial amounts of EpiC, such as green tea, cocoa, apple peel and black tea, can enhance memory formation in *Lymnaea* [7]. Exposed to pond water containing green tea or pure cocoa powder in concentration comparable to human consumption level (approximately 1 g/day) during training, the memory enhancement was comparable to that elicited by pure EpiC experiments [7].

Interestingly, black tea does not only enhance LTM formation but suppresses LTM formation in *Lymnaea* [61]. Black tea is made from the same plant as green-tea through an oxidation process called "fermentation" and becomes stronger in flavor than green tea. However, the content of EpiC in black tea (0.49 mg/100 g) reduces compared with green tea (6.16 mg/100 g) [3]. Black tea substantially contains more other flava-3-nols, thearubigins and theaflavins, than green tea [3, 62]. These flava-3-nols are formed during the fermentation reaction in black tea. As far as we know, no studies have investigated direct effects of these flava-3-nols on memory formation. However, it is reported that intake of theaflavins is associated with long-term language and verbal memory in human [63]. In another study, theaflavins are reported to improve memory impairment [64, 65]. Thus, thearubigins and theaflavins may be not candidate substances for blocking memory formation in black tea.

Another component of black tea, caffeine, can inhibit cognitive function. In drosophila, caffeine reduces the performance for light aversive conditioning [66]. However, both green tea and black tea contain a same amount of caffeine. It is possible that L-theanine included in green tea in comparatively large amounts is thought to balance the effects of caffeine. The combination of these two substances may be synergistic, as one study found that people who ingested L-theanine and caffeine together had better attention than when either was used alone [67, 68]. Therefore, investigating catechin-rich foods is difficult to permit a full understanding of the specific effect of these phytochemicals.

5. Rescue effect of epicatechin on stress-impaired memory

Green tea-derived catechins do not only enhance memory formation, but also rescue impaired cognitive functions due to environmental stressors. Catechin-rich foods have been considered to improve various aspects of cognitive functions in rodents and humans, and some reports suggest that it has positive effects on mild cognitive impairment [69–71]. EpiC administration improves spatial memory in mice via an increase in cerebral angiogenesis or a direct effect on neural elements [8]. In *Lymnaea*, there are some reports for the recovery effect of EpiC on impaired function by environmental stressor [5, 38].

Most of freshwater mollusks including *Lymnaea* are dependent on calcium intake directly from the environment through their skin [72] and exhibit reduced shell growth in the environment containing less than 20 mg/L calcium [73, 74]. It is considered that this level of calcium acts as a stressor on the snail. Following 1 h exposure to a low calcium environment, *Lymnaea* was not able to form LTM, although it still had an ability of ITM [75]. Following a 2 h-TS in EpiC-supplemented low-calcium pond water, snails persist a decrease of respiratory behavior both 24 hours and 72 hours after the training [5]. In addition, memory formation of the training in EpiC-suppelmented pondwater was not diminished by the combination of a low- calcium pond water environment and 1 hour of crowding immediately prior to operant conditioning training, which blocks all forms of memory (short-term, intermediate-term and long-term memory) in *Lymnaea* [38]. These results suggest that EpiC reverses an imposed memory deficit by exposure to memory 'unfriendly' stress.

6. Enhancing effect of epicatechin on memory formation by classical conditioning of feeding behavior

Lymnaea can be classically, as well as operantly, conditioned [28, 29]. Conditioned taste aversion (CTA) is a classical conditioning, which is based on pairing sucrose as a conditioned stimulus (CS) with an aversive chemical unconditioned stimulus (UCS) such as KCl, which inhibits feeding and evokes a withdrawal response. After this procedure, trained snails show a significantly weaker feeding response to sucrose than controls. We here introduced the enhancing effect of EpiC on LTM formed by CTA. In the previous study, we showed that EpiC increases the persistence of LTM as mentioned below [76].

CTA training procedure we performed is briefly as follows. Adult snails randomly chosen were food deprived for 24 hours before being subjected to CTA training. Snails were then immersed in an appetitive solution (10 mM sucrose) for 15 s. Then, the sucrose solution was quickly replaced with distilled water, and the feeding response (i.e. number of bites) was measured in distilled water for 5 minutes (pre-test). Ten minutes after the pre-test, CTA training was performed. In CTA training, snails were immersed for 15 s in 10 mM sucrose, which were immediately immersed for 15 s in 10 mM KCl solution (i.e. the UCS). The UCS

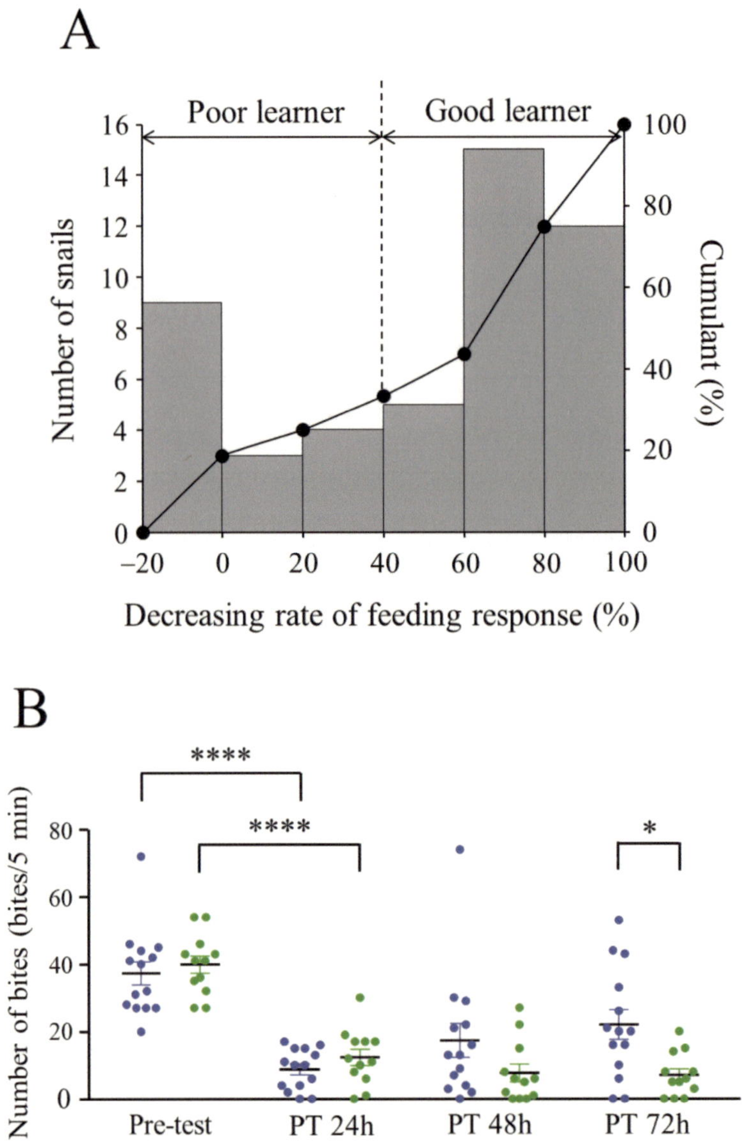

Figure 4.
*Change in feeding response after conditioned taste aversion (CTA) training. (A) Histogram showing the decreasing rate of the feeding response at the 24 h post-test n = 48). The cumulant (the ratio of the cumulative number of snails at each rate, from low to high, out of the total number, n = 48; black circles) is also shown for reference. Snails that showed at least a 40% decrease in the number of bites were defined as good learners, while poor learners were defined as snails whose post-test scores decreased by less than 40%. (B) Feeding responses in the pre- and post-tests for good learners trained without Epi (control, blue circles, n = 14) and with Epi (green circles, n = 12). ****P < 0.0001, *P < 0.05. These figures are reproduced from [76] with permission.*

.inhibits the feeding response. After the UCS was presented, snails were immersed either in distilled water (control) or EpiC solution (15 mg/L) for 9.5 minutes. This procedure was repeated 5 times. After CTA training, snails were kept in distilled water for 24, 48 or 72 hours and then the post-test was performed, which was exactly the same as the pre-test. By comparing the number of bites in the pre-test with that in the 24 h post-test, we determined whether the snail was a 'good' learner or a 'poor' learner.

Figure 4A shows a histogram of the decreasing rate of the feeding response in the 24 h post-test (i.e. 24 hours after training). The decreasing rate was measured for 26 snails trained without EpiC and 22 snails trained with EpiC, and the data from all snails (n = 48) are combined in the histogram. As shown in **Figure 4A**, from their responses, the snails were roughly divided into two groups: 'good' and 'poor' learners. Good learners were defined as snails that showed at least a 40% decrease in the number of bites in the 24 h post-test compared with that in the pre-test. Thus, poor learners were defined as snails whose post-test scores decrease by less than 40%. In this data, 32 of the 48 snails (i.e. 67%) were classified as good learners.

For the good learners, we statistically analyzed on the data presented in **Figure 4B** (control group, n = 14; EpiC group, n = 12). In both groups, snails showed a significant decrease in the number of bites in the 24 h post-test (control group, 37.3 ± 3.5 to 8.8 ± 1.6 bites per 5 min, $P < 0.0001$; EpiC group: 39.9 ± 2.6 to 12.3 ± 2.4 bites per 5 min, $P < 0.0001$). Thus, placing snails in the EpiC solution in CTA training did not alter their 24 h memory performance.

We next compared the post-test scores between two groups (control group versus EpiC group) at 24, 48 and 72 hours after CTA training. A statistical analyze showed that there was a significant difference in the memory scores at 72 hours between the two groups (control group versus EpiC group, $P < 0.05$) while there was no significant difference in the 24 h and 48 h post-tests. Thus, we concluded that exposing snails to EpiC solution resulted in significantly longer memory persistence.

An identified spontaneously active pair of neurons, the cerebral giant cells (CGCs), has been shown to both modulate the neuronal network underling the feeding behavior and be necessary for LTM and its retrieval following CTA training (**Figure 4**) [31, 32]. Therefore, a possible mechanism underlying the significant effect of EpiC on LTM persistence is an alteration in CGC activity. Our data supported this possibility [76]. Additionally, our data suggested that a GABAergic neuron may play a significant role in mediating CTA-LTM and the EpiC effect on the CGC may involve a GABAergic neuron. For example, the GABA sensitivity of a neuron (maybe the CGC itself) might be enhanced in good learners or in snails exposed to EpiC.

7. Conclusion

Studies described in this chapter have provided valuable information on a possibility of EpiC-rich foods contributing to cognition ability in *Lymnaea*. In addition, animal models in *Lymnaea* contribute to new evidences for the generality of mechanisms for the effects of EpiC on learning and memory formation, across learning paradigms (e.g. classical or operant conditioning).

These studies suggest that EpiC has not only antioxidant properties but also targets molecules (e.g. specific receptors) directly to affect the signaling pathway. Then, the results may yield the basis of future studies to elucidate how EpiC enhances LTM formation of classical and operant conditioning in *Lymnaea*.

Author details

Yoshimasa Komatsuzaki[1], Ayaka Itoh[2] and Minoru Saito[2]*

1 Department of Physics, College of Science and Technology, Nihon University, Tokyo, Japan

2 Department of Correlative Study in Physics and Chemistry, Graduate School of Integrated Basic Sciences, Nihon University, Tokyo, Japan

*Address all correspondence to: saitou.minoru79@nihon-u.ac.jp

References

[1] Gupta AK, Daigle D. Sinecatechins 10% ointment: a green tea extract for the treatment of external genital warts. Skin Therapy Lett. 2015;20:6-8. DOI:

[2] Spencer JP, Abd El Mohsen MM, Minihane AM, Mathers JC. Biomarkers of the intake of dietary polyphenols: strengths, limitations and application in nutrition research. Br J Nutr. 2008; 99:12-22. DOI: 10.1017/S000711450 7798938

[3] Bhagwat SA, Haytowitz DB, Wasswa-Kintu SI, Pehrsson PR. Process of formulating USDA's Expanded Flavonoid Database for the Assessment of Dietary intakes: a new tool for epidemiological research. Brit J Nutr. 2015;114:472-80. DOI: 10.1017/ S0007114515001580

[4] Fruson L, Dalesman S, Lukowiak K. A flavonol present in cocoa [(−) epicatechin] enhances snail memory. J Exp Biol. 2012;215:3566-76. DOI: 10.1242/jeb.070300

[5] Knezevic B, Lukowiak K. The flavonol epicatechin reverses the suppressive effects of a stressor on long-term memory formation. Journal of Experimental Biology. 2014;217:4004-9. DOI: 10.1242/ jeb.110726

[6] Li Q, Zhao HF, Zhang ZF, Liu ZG, Pei XR, Wang JB, et al. Long-term administration of green tea catechins prevents age-related spatial learning and memory decline in C57BL/6 J mice by regulating hippocampal cyclic amp-response element binding protein signaling cascade. Neuroscience. 2009;159:1208-15. DOI: 10.1016/j. neuroscience.2009.02.008

[7] Swinton E, de Freitas E, Swinton C, Shymansky T, Hiles E, Zhang J, et al. Green tea and cocoa enhance cognition in Lymnaea. Commun Integr Biol. 2018;11:e1434390. DOI: 10.1080/19420889.2018.1434390

[8] van Praag H, Lucero MJ, Yeo GW, Stecker K, Heivand N, Zhao C, et al. Plant-derived flavanol (−)epicatechin enhances angiogenesis and retention of spatial memory in mice. J Neurosci. 2007;27:5869-78. DOI: 10.1523/ JNEUROSCI.0914-07.2007

[9] Abel T, Kandel E. Positive and negative regulatory mechanisms that mediate long-term memory storage. Brain Res Brain Res Rev. 1998;26:360-78. DOI: 10.1016/s0165-0173(97) 00050-7

[10] Bailey CH, Kandel ER. Structural changes accompanying memory storage. Annu Rev Physiol. 1993;55:397-426. DOI: 10.1146/annurev.ph.55.030193. 002145

[11] Gelperin A, Kleinfeld D, Denk W, Cooke IR. Oscillations and gaseous oxides in invertebrate olfaction. J Neurobiol. 1996;30:110-22. DOI: 10.1002/(SICI)1097-4695(199605) 30:1<110::AID-NEU10>3.0.CO;2-Q

[12] Ito I, Kimura T, Suzuki H, Sekiguchi T, Ito E. Effects of electrical stimulation of the tentacular digits of a slug upon the frequency of electrical oscillations in the procerebral lobe. Brain Res. 1999;815:121-5. DOI: 10.1016/ s0006-8993(98)01115-9

[13] Kimura T, Iwama A, Sekiguchi T. Contributions of superior and inferior tentacles to learned food-avoidance behavior in *Limax marginatus*. Zool Sci. 1999;16:595-602. DOI: 10.2108/ zsj.16.595

[14] Matzel LD, Talk AC, Muzzio IA, Rogers RF. Ubiquitous molecular substrates for associative learning and activity-dependent neuronal

facilitation. Rev Neurosci. 1998;9:129-67. DOI: 10.1515/revneuro.1998.9.3.129

[15] Lukowiak K, Cotter R, Westly J, Ringseis E, Spencer G. Long-term memory of an operantly conditioned respiratory behaviour pattern in lymnaea stagnalis. J Exp Biol. 1998;201 (Pt 6):877-82. DOI:

[16] Lukowiak K, Ringseis E, Spencer G, Wildering W, Syed N. Operant conditioning of aerial respiratory behaviour in *Lymnaea stagnalis*. J Exp Biol. 1996;199:683-91. DOI:

[17] Lukowiak K, Sangha S, McComb C, Varshney N, Rosenegger D, Sadamoto H, et al. Associative learning and memory in *Lymnaea stagnalis*: how well do they remember? J Exp Biol. 2003;206:2097-103. DOI: 10.1242/jeb.00374

[18] Scheibenstock A, Krygier D, Haque Z, Syed N, Lukowiak K. The Soma of RPeD1 must be present for long-term memory formation of associative learning in Lymnaea. J Neurophysiol. 2002;88:1584-91. DOI: 10.1152/jn.2002.88.4.1584

[19] Spencer GE, Syed NI, Lukowiak K. Neural changes after operant conditioning of the aerial respiratory behavior in *Lymnaea stagnalis*. J Neurosci. 1999;19:1836-43. DOI:

[20] Lukowiak K, Sangha S, Scheibenstock A, Parvez K, McComb C, Rosenegger D, et al. A molluscan model system in the search for the engram. J Physiol Paris. 2003;97:69-76. DOI: 10.1016/j.jphysparis.2003.10.008

[21] Braun MH, Lukowiak K. Intermediate and long-term memory are different at the neuronal level in *Lymnaea stagnalis* (L.). Neurobiol Learn Mem. 2011;96:403-16. DOI: 10.1016/j.nlm.2011.06.016

[22] Sangha S, Scheibenstock A, Morrow R, Lukowiak K. Extinction requires new RNA and protein synthesis and the soma of the cell right pedal dorsal 1 in *Lymnaea stagnalis*. J Neurosci. 2003;23:9842-51. DOI:

[23] Syed NI, Bulloch AG, Lukowiak K. In vitro reconstruction of the respiratory central pattern generator of the mollusk Lymnaea. Science. 1990;250:282-5. DOI: 10.1126/science.2218532

[24] Syed NI, Ridgway RL, Lukowiak K, Bulloch AG. Transplantation and functional integration of an identified respiratory interneuron in *Lymnaea stagnalis*. Neuron. 1992;8:767-74. DOI: 10.1016/0896-6273(92)90097-w

[25] Sangha S, Scheibenstock A, Lukowiak K. Reconsolidation of a long-term memory in Lymnaea requires new protein and RNA synthesis and the soma of right pedal dorsal 1. J Neurosci. 2003;23:8034-40. DOI:

[26] McComb C, Rosenegger D, Varshney N, Kwok HY, Lukowiak K. Operant conditioning of an in vitro CNS-pneumostome preparation of Lymnaea. Neurobiol Learn Mem. 2005;84:9-24. DOI: 10.1016/j.nlm.2005.02.002

[27] Rosenegger D, Parvez K, Lukowiak K. Enhancing memory formation by altering protein phosphorylation balance. Neurobiol Learn Mem. 2008;90:544-52. DOI: 10.1016/j.nlm.2008.06.005

[28] Ito E, Kojima S, Lukowiak K, Sakakibara M. From likes to dislikes: conditioned taste aversion in the great pond snail (*Lymnaea stagnalis*). Can J Zool. 2013;91:405-12. DOI: 10.1139/cjz-2012-0292

[29] Lukowiak K, Sunada H, Teskey M, Lukowiak K, Dalesman S. Environmentally relevant stressors alter memory formation in the pond snail Lymnaea. J Exp Biol. 2014;217:76-83. DOI: 10.1242/jeb.089441

[30] Kojima S, Yamanaka M, Fujito Y, Ito E. Differential neuroethological effects of aversive and appetitive reinforcing stimuli on associative learning in *Lymnaea stagnalis*. Zool Sci. 1996;13:803-12. DOI: DOI 10.2108/zsj.13.803

[31] Benjamin PR. Distributed network organization underlying feeding behavior in the mollusk Lymnaea. Neural Syst Circuits. 2012;2:4. DOI: 10.1186/2042-1001-2-4

[32] Sunada H, Horikoshi T, Lukowiak K, Sakakibara M. Increase in excitability of RPeD11 results in memory enhancement of juvenile and adult *Lymnaea stagnalis* by predator-induced stress. Neurobiology of Learning and Memory. 2010;94:269-77. DOI: 10.1016/j.nlm.2010.06.005

[33] Ito E, Otsuka E, Hama N, Aonuma H, Okada R, Hatakeyama D, et al. Memory trace in feeding neural circuitry underlying conditioned taste aversion in Lymnaea. PLoS One. 2012;7:e43151. DOI: 10.1371/journal.pone.0043151

[34] Kojima S, Hosono T, Fujito Y, Ito E. Optical detection of neuromodulatory effects of conditioned taste aversion in the pond snail *Lymnaea stagnalis*. J Neurobiol. 2001;49:118-28. DOI: 10.1002/neu.1069

[35] Kojima S, Nanakamura H, Nagayama S, Fujito Y, Ito E. Enhancement of an inhibitory input to the feeding central pattern generator in *Lymnaea stagnalis* during conditioned taste-aversion learning. Neurosci Lett. 1997;230:179-82. DOI: 10.1016/s0304-3940(97)00507-7

[36] Yeoman MS, Brierley MJ, Benjamin PR. Central pattern generator interneurons are targets for the modulatory serotonergic cerebral giant cells in the feeding system of Lymnaea. J Neurophysiol. 1996;75:11-25. DOI: 10.1152/jn.1996.75.1.11

[37] Parvez K, Moisseev V, Lukowiak K. A context-specific single contingent-reinforcing stimulus boosts intermediate-term memory into long-term memory. Eur J Neurosci. 2006;24:606-16. DOI: 10.1111/j.1460-9568.2006.04952.x

[38] Knezevic B, Komatsuzaki Y, de Freitas E, Lukowiak K. A flavanoid component of chocolate quickly reverses an imposed memory deficit. Journal of Experimental Biology. 2016;219:816-23. DOI: 10.1242/jeb.130765

[39] Sangha S, McComb C, Lukowiak K. Forgetting and the extension of memory in Lymnaea. J Exp Biol. 2003;206:71-7. DOI: 10.1242/jeb.00061

[40] Sangha S, Scheibenstock A, Martens K, Varshney N, Cooke R, Lukowiak K. Impairing forgetting by preventing new learning and memory. Behav Neurosci. 2005;119:787-96. DOI: 10.1037/0735-7044.119.3.787

[41] Abd El Mohsen MM, Kuhnle G, Rechner AR, Schroeter H, Rose S, Jenner P, et al. Uptake and metabolism of epicatechin and its access to the brain after oral ingestion. Free Radic Biol Med. 2002;33:1693-702. DOI: 10.1016/s0891-5849(02)01137-1

[42] Fernell M, Swinton C, Lukowiak K. Epicatechin, a component of dark chocolate, enhances memory formation if applied during the memory consolidation period. Commun Integr Biol. 2016;9:e1205772. DOI: 10.1080/19420889.2016.1205772

[43] Lukowiak K, Adatia N, Krygier D, Syed N. Operant conditioning in Lymnaea: evidence for intermediate- and long-term memory. Learn Mem. 2000;7:140-50. DOI: 10.1101/lm.7.3.140

[44] Sangha S, Morrow R, Smyth K, Cooke R, Lukowiak K. Cooling blocks ITM and LTM formation and preserves memory. Neurobiol Learn Mem.

2003;80:130-9. DOI: 10.1016/s1074-7427(03)00065-0

[45] Takahashi T, Takigami S, Sunada H, Lukowiak K, Sakakibara M. Critical period of memory enhancement during taste avoidance conditioning in *Lymnaea stagnalis*. PLoS One. 2013;8:e75276. DOI: 10.1371/journal.pone.0075276

[46] Nehlig A. The neuroprotective effects of cocoa flavanol and its influence on cognitive performance. Br J Clin Pharmacol. 2013;75:716-27. DOI: 10.1111/j.1365-2125.2012.04378.x

[47] Shi M, Nie Y, Zheng XQ, Lu JL, Liang YR, Ye JH. Ultraviolet B (UVB) Photosensitivities of Tea Catechins and the Relevant Chemical Conversions. Molecules. 2016;21. DOI: 10.3390/molecules21101345

[48] Sangha S, Scheibenstock A, McComb C, Lukowiak K. Intermediate and long-term memories of associative learning are differentially affected by transcription versus translation blockers in Lymnaea. J Exp Biol. 2003;206:1605-13. DOI: 10.1242/jeb.00301

[49] Lynch MA. Long-term potentiation and memory. Physiol Rev. 2004;84:87-136. DOI: 10.1152/physrev.00014.2003

[50] Il-Han J, Janes T, Lukowiak K. The role of serotonin in the enhancement of long-term memory resulting from predator detection in Lymnaea. J Exp Biol. 2010;213:3603-14. DOI: 10.1242/jeb.048256

[51] Moreno-Ulloa A, Mendez-Luna D, Beltran-Partida E, Castillo C, Guevara G, Ramirez-Sanchez I, et al. The effects of (−)-epicatechin on endothelial cells involve the G protein-coupled estrogen receptor (GPER). Pharmacol Res. 2015;100:309-20. DOI: 10.1016/j.phrs.2015.08.014

[52] Sadamoto H, Sato H, Kobayashi S, Murakami J, Aonuma H, Ando H, et al. CREB in the pond snail *Lymnaea stagnalis*: cloning, gene expression, and function in identifiable neurons of the central nervous system. J Neurobiol. 2004;58:455-66. DOI: 10.1002/neu.10296

[53] Nichols M, Zhang J, Polster BM, Elustondo PA, Thirumaran A, Pavlov EV, et al. Synergistic neuroprotection by epicatechin and quercetin: Activation of convergent mitochondrial signaling pathways. Neuroscience. 2015;308:75-94. DOI: 10.1016/j.neuroscience.2015.09.012

[54] Granado-Serrano AB, Angeles Martin M, Goya L, Bravo L, Ramos S. Time-course regulation of survival pathways by epicatechin on HepG2 cells. J Nutr Biochem. 2009;20:115-24. DOI: 10.1016/j.jnutbio.2007.12.006

[55] Rosenegger D, Lukowiak K. The participation of NMDA receptors, PKC, and MAPK in the formation of memory following operant conditioning in Lymnaea. Mol Brain. 2010;3:24. DOI: 10.1186/1756-6606-3-24

[56] Shirakami Y, Shimizu M. Possible Mechanisms of Green Tea and Its Constituents against Cancer. Molecules. 2018;23. DOI: 10.3390/molecules23092284

[57] Sunada H, Riaz H, de Freitas E, Lukowiak K, Swinton C, Swinton E, et al. Heat stress enhances LTM formation in Lymnaea: role of HSPs and DNA methylation. J Exp Biol. 2016;219:1337-45. DOI: 10.1242/jeb.134296

[58] Stringer TP, Guerrieri D, Vivar C, van Praag H. Plant-derived flavanol (−) epicatechin mitigates anxiety in association with elevated hippocampal monoamine and BDNF levels, but does not influence pattern separation in mice. Transl Psychiatry. 2015;5:e493. DOI: 10.1038/tp.2014.135

[59] Moroz LL, Winlow W, Turner RW, Bulloch AG, Lukowiak K, Syed NI.

Nitric oxide synthase-immunoreactive cells in the CNS and periphery of Lymnaea. Neuroreport. 1994;5:1277-80. DOI: 10.1097/00001756-1994 06020-00031

[60] Orr MV, Lukowiak K. Electrophysiological and behavioral evidence demonstrating that predator detection alters adaptive behaviors in the snail Lymnaea. J Neurosci. 2008;28:2726-34. DOI: 10.1523/ JNEUROSCI.5132-07.2008

[61] Zhang J, de Freitas E, Lukowiak K. Black tea differs from green tea: it suppresses long-term memory formation in Lymnaea. Commun Integr Biol. 2018;11:1-4. DOI: 10.1080/ 19420889.2018.1491245

[62] Peluso I, Serafini M. Antioxidants from black and green tea: from dietary modulation of oxidative stress to pharmacological mechanisms. Br J Pharmacol. 2017;174:1195-208. DOI: 10.1111/bph.13649

[63] Kesse-Guyot E, Fezeu L, Andreeva VA, Touvier M, Scalbert A, Hercberg S, et al. Total and specific polyphenol intakes in midlife are associated with cognitive function measured 13 years later. J Nutr. 2012;142:76-83. DOI: 10.3945/ jn.111.144428

[64] Ano Y, Ohya R, Kita M, Taniguchi Y, Kondo K. Theaflavins Improve Memory Impairment and Depression-Like Behavior by Regulating Microglial Activation. Molecules. 2019;24. DOI: 10.3390/molecules24030467

[65] Murphy KJ, Walker KM, Dyer KA, Bryan J. Estimation of daily intake of flavonoids and major food sources in middle-aged Australian men and women. Nutr Res. 2019;61:64-81. DOI: 10.1016/j.nutres.2018.10.006

[66] Folkers E, Spatz HC. Visual Learning-Performance of Drosophila-Melanogaster Is Altered by Neuropharmaca Affecting Phosphodiesterase Activity and Acetylcholine Transmission. J Insect Physiol. 1984;30:957-65. DOI: Doi 10.1016/0022-1910(84)90074-X

[67] Boros K, Jedlinszki N, Csupor D. Theanine and Caffeine Content of Infusions Prepared from Commercial Tea Samples. Pharmacogn Mag. 2016;12:75-9. DOI: 10.4103/ 0973-1296.176061

[68] Kelly SP, Gomez-Ramirez M, Montesi JL, Foxe JJ. L-theanine and caffeine in combination affect human cognition as evidenced by oscillatory alpha-band activity and attention task performance. J Nutr. 2008;138:1572S-7S. DOI: 10.1093/jn/138.8.1572S

[69] Park SK, Jung IC, Lee WK, Lee YS, Park HK, Go HJ, et al. A combination of green tea extract and l-theanine improves memory and attention in subjects with mild cognitive impairment: a double-blind placebo-controlled study. J Med Food. 2011;14:334-43. DOI: 10.1089/ jmf.2009.1374

[70] Mancini E, Beglinger C, Drewe J, Zanchi D, Lang UE, Borgwardt S. Green tea effects on cognition, mood and human brain function: A systematic review. Phytomedicine. 2017;34:26-37. DOI: 10.1016/j.phymed.2017.07.008

[71] Yamada T, Yamada Y, Okano Y, Terashima T, Yokogoshi H. Anxiolytic effects of short- and long-term administration of cacao mass on rat elevated T-maze test. J Nutr Biochem. 2009;20:948-55. DOI: 10.1016/j. jnutbio.2008.08.007

[72] Van der Borght O, Van Puymbroeck S. Calcium metabolism in a freshwater mollusc: quantitative importance of water and food as supply for calcium during growth. Nature. 1966;210:791-3. DOI: 10.1038/210791a0

[73] Rundle SD, Spicer JI, Coleman RA, Vosper J, Soane J. Environmental calcium modifies induced defences in snails. Proc Biol Sci. 2004;271 Suppl 3:S67-70. DOI: 10.1098/rsbl.2003.0106

[74] Greenaway P. Calcium regulation in the freshwater mollusc, Limnaea stagnalis (L.) (Gastropoda: Pulmonata). I. The effect of internal and external calcium concentration. J Exp Biol. 1971;54:199-214. DOI:

[75] Dalesman S, Braun MH, Lukowiak K. Low environmental calcium blocks long-term memory formation in a freshwater pulmonate snail. Neurobiol Learn Mem. 2011;95:393-403. DOI: 10.1016/j.nlm.2010.11.017

[76] Itoh A, Komatsuzaki Y, Lukowiak K, Saito M. Epicatechin increases the persistence of long-term memory formed by conditioned taste aversion in Lymnaea. J Exp Biol. 2021;224. DOI: 10.1242/jeb.238055

Chapter 3

Seasonal Variations of Densities of *Biomphalaria pfeifferi*, the Intermediate Host of *Schistosoma mansoni* Parasite at the North of Senegal

Sidy Bakhoum, Christopher J.E. Haggerty, Cheikh Tidiane Ba, Nicolas Jouanard, Gilles Riveau and Jason Robert Rohr

Abstract

Schistosomiasis is becoming more persistent because of the widespread distribution of intermediate host snails in several regions of Africa, including Senegal. The intermediate snail host of the human intestinal schistosome is *Biomphalaria pfeifferi* and is permanently present in northern Senegal because of the presence of the abundant freshwater habitat throughout the year. Here, we observed the seasonal variation in *B. pfeifferi* abundance in the Saint-louis region at the North of Senegal in West Africa. We performed snail and environmental parameter sampling across two different seasons described for Senegal: a dry season that runs roughly from mid-October to mid-June and a rainy season that spans approximately from late June to early October. We also split the dry season into two categories representing periods of time when water temperatures were either decreasing (dry1) or increasing (dry2). We used regression analyses to model snail density across the seasons and investigated which environmental variables influenced snail abundance. Results suggested that snails were more abundant and peaked during the rainy season, which lowest abundances during the dry season when temperatures were declining. The above seasonal variations of snail density were positively linked to the environmental drivers including periphyton (food resource for snails), aquatic vegetation abundance, water temperature and dissolved oxygen and negatively to both pH and water conductivity. Our findings may be useful for snail control efforts by targeting specific periods and/or site conditions when snail abundances are greatest.

Keywords: schistosomiasis, intermediate hosts, snails, seasonal variation, *Biomphalaria pfeifferi*, *Schistosoma mansoni*

1. Introduction

To completely understand many infectious disease systems, it will become increasingly important to understand how seasonality affects multiple processes, including (but not limited to): host behavior, reproduction, survival in the environment [1]. Seasonal variations in temperature, rainfall and resource availability are ubiquitous and can exert strong pressures on host population dynamics.

Infectious diseases provide some of the best-studied examples of the role of seasonality in shaping population fluctuations [1]. Predicting disease dynamics requires an understanding of the traits of hosts across seasons.

Schistosomiasis remains a significant health burden in many areas of the world [2]. Intermediate hosts snails of human schistosomes release parasites that cause human schistosomiasis, are strongly driven by environmental factors [3–5]. The transformation of ecosystems in the river delta Senegal has created favorable biotopes to the development of intermediate host snails of human schistosomes [3], and schistosomiasis rates have increased from historic levels. Thus, limiting or controlling snail populations is an important step in disease control. Yet, consideration of the relative influence of seasonality on the environmental factors that influence snail host populations remains needed.

In this study, we investigated how snail host abundance varies at water access sites in North Senegal, West Africa, a location where human schistosomiasis prevalence is among the highest globally. We conducted our study in Senegal River at four sites used for water access in three villages. We performed biweekly monitoring of *Biomphalaria pfeifferi* across two different seasons in Senegal: a dry season that runs roughly from mid-October to mid-June and a rainy season spanning approximately from late June to early October. Further, we sampled two periods of water temperature T (°C): Dry_1 (periods of decreasing temperature ($29.4 \geq T (Dry_1) \geq 15.4$) and Dry_2 (periods of increasing temperature ($15.5 < T (Dry_2) \leq 32.5$)). We conducted a total of $n = 21$ visits to survey density of hosts and the environmental parameters in the field (as described below) for one year from October 2019 to October 2020.

2. Methods

2.1 Study sites

We sampled intermediate hosts biweekly at four water access points across three villages in Senegal, West Africa: Kaban (KA: 16° 3.338 N - 16°24.133 O) Minguegne (ME 1: 16°01.055' N - 16°21.397' O and ME2:16°01.090' N - 16°21.369 O) and

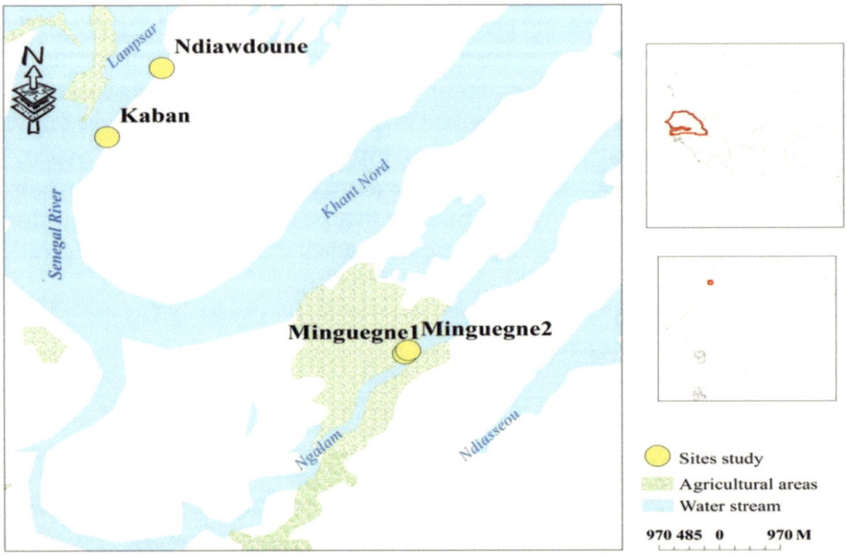

Figure 1.
Location of the study sites along the Senegal River in West Africa.

Ndiawdoune (NW: 16°4.075' N - 16°23.635' O). Kaban village is bordered by the Senegal river, Minguegne is bordered to the Ngalam outlet Senegal River in the "Trois marigots" zone and Ndiawdoune is bordered by the Lampsar River (**Figure 1**).

2.2 Environmental factors driving host snail abundance

We recorded dissolved oxygen (DO), pH, water conductivity and water temperature using a YSI Professional Plus handheld multiparameter meter. We also recorded periphyton fluorescence using an Aquapen AP 100-C handheld flourometer. Periphyton was collected from a study site during snail sampling (see below) by wading into the water and cutting a stem of *Typha* spp. vegetation at the water surface and again at a depth of 10 cm below the water. The 10 cm section of Typha was taken to the lab and its surface scrubbed using a toothbrush. We then washed all algae off the Typha and brush using deionized water into a 50 mL falcon tube and filled all samples to a standardized volume of 50 mL. Chlorphyll a was then quantified by filling a 1 mL cuvette and recording Ft values using the Aquapen. We recorded the surface area of *Typha* that was sampled and used it to standardize the periphyton flouresence values based on sampling area.

2.3 Snail sampling

We used the snail sampling method described in [6]. At each water access point, we conducted 10 1-m sweeps with a 2.5-mm mesh aquatic dipnet at random sampling points. Any aquatic plants in the dipnet were placed into a wash pale with water, shaken vigorously to remove snails, and examined for any attached snails before weighing the vegetation mass using a spring scale. We recorded the number of *Biomphalaria pfeiferii* snails (**Figure 2**) captured per sweep.

2.4 Statistical analysis

We conducted our data analysis using R software version string *R version 4.0.5 (2021-2103-31)*. We applied a forest plot in the *Forest model* package to a linear model to assess the variation of snail's density across the seasons. In forest plots, a line of no effect (at 0) marks the point where there is no clear difference between the variables. If the 95% confidence intervals (CIs) do not cross the line of no effect, then the result is significant (p-value <0.05). To elucidate actions between

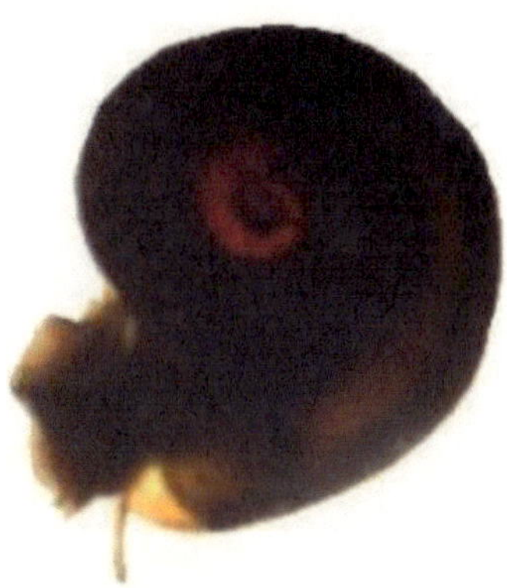

Figure 2.
Biomphalaria pfeifferi *snail species of the current study. This is the only known intermediate host of* Schistosoma mansoni, *the parasite of the intestinal schistosomiasis in Senegal.*

environmental parameters and hosts density, we applied Pearson's correlation, which gives a measure of the strength of a linear association between two variables. We also determined the *p-value* or probability that we would have found the current result if the correlation coefficient *r* were in fact zero (null hypothesis). If this probability is lower than the conventional 5% ($p < 0.05$) the correlation coefficient is called statistically significant. We used linear model in *ggplot2* package that utilizes a *lm* function to assess the significance of linear predictor-response relationships between hosts and environmental drivers.

3. Results

3.1 Seasonal variations and environmental parameters driving the densities of the snail's hosts

We collected a total *895 Biomphalaria pfeifferi*, including 83 in Dry1 (9.27%), 192 in Dry2 (21.45%), and 620 in rainy (69.28%) seasons (**Table 1**). A factor for season in our forest plot analyses was a significant predictor of snail abundance (**Table 2**, *p-value* = 0.003), with both greater mean of total density (*N*) and maximum snail density per visit occurring in the rainy season (**Table 1**). In contrast, all measures of snail abundance were lowest during Dry1 season, whereas a medium density occurred in Dry2 season (**Figure 3**). Our forest plot analyses suggested that snail abundance in the rainy season was significantly different from season Dry1, whereas season Dry2 had lower but not significantly lower snail abundance compared to the rainy season (**Figure 2**; *p-value* = 0.181).

During Dry1 season, when snail abundance was lowest, no environmental parameters had a significant association to *Biomphalaria* abundance (**Table 1**;

		Mean	Max
Seasons	Dry1 (*N = 83*)	3.458	18.000
	Dry2 (*N = 192*)	9.6	33.0
	Rainy (*N = 620*)	15.5	87.0
Total *(N = 895)*		10.65	87.0

Table 1.
Total, mean and maximum of the total and seasonal densities of Biomphalaria pfeifferi.

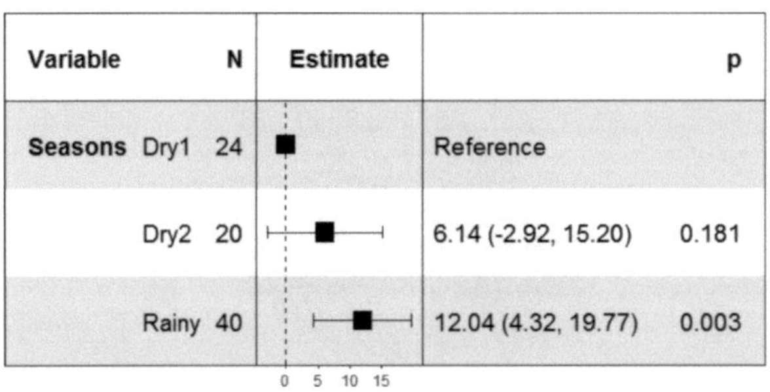

Table 2.
Forest plot analyses of Biomphalaria pfeifferi *abundance across seasons.*

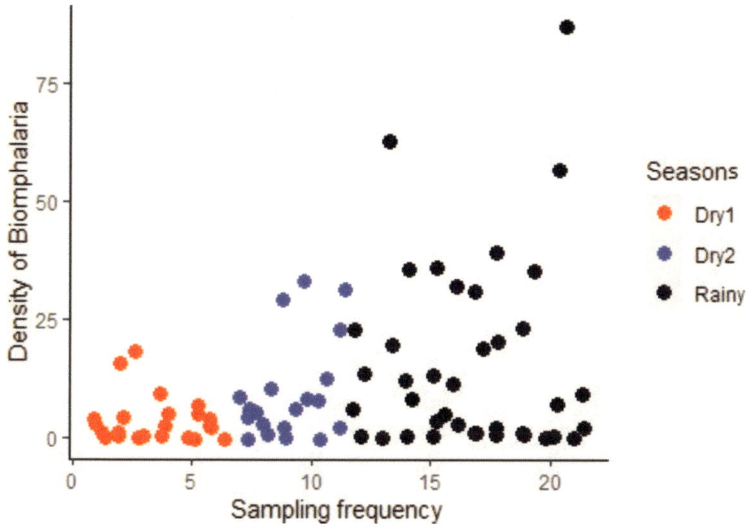

Figure 3.
Plot showing the distribution of Biomphalaria pfeifferi *across the seasons. The greatest density was collected during the rainy season (with a peak >75) while the weakest was found during the Dry1 season.*

Environmental parameters	Density of *Biomphalaria pfeifferi*		
	Dry1	**Dry2**	**Rainy**
Temperature (°C)	0.991	0.04*	0.549
Conductivity (µS/cm)	0.111	0.02*	0.08
DO (mg/l)	0.447	0.006**	0.04*
Periphyton	0.431	< 0.001***	0.003**
Mass vegetation (g)	0.452	0.007**	0.05
pH	0.429	0.03*	0.388

**P ≤ 0.05*
***P ≤ 0.01*
****P ≤ 0.001*
If P > 0.05, there is no significance (no asterisk).

Table 3.
Linear regressions between environmental factors and density of Biomphalaria pfeifferi for each season of snail sampling.

Figure 3; *p-value* > 0.05). In contrast, several environmental factors were significantly correlated to the snail host abundance in season Dry_2 (**Table 3**). Conductivity and pH were negatively ($r < 0$) and significantly associated to hosts abundance whereas other drivers are positively ($r > 0$) correlated in season Dry_2. The maximum water temperature (32.5° C) was obtained during rainy season. Water temperature was positively related to snail abundance, exerted a significant effect only in the season Dry_2 (**Table 3** and **Figure 3**). There was no significant effect of temperature on snail densities in the rainy season (p-value = 0.549). Only the dissolved oxygen (p-value = 0.04) and periphyton (p-value = 0.003) were significantly correlated to the density of hosts in the rainy season. Periphyton was a significant positive predictor of snail abundances in both the Dry_2 and rainy seasons (**Figure 4**).

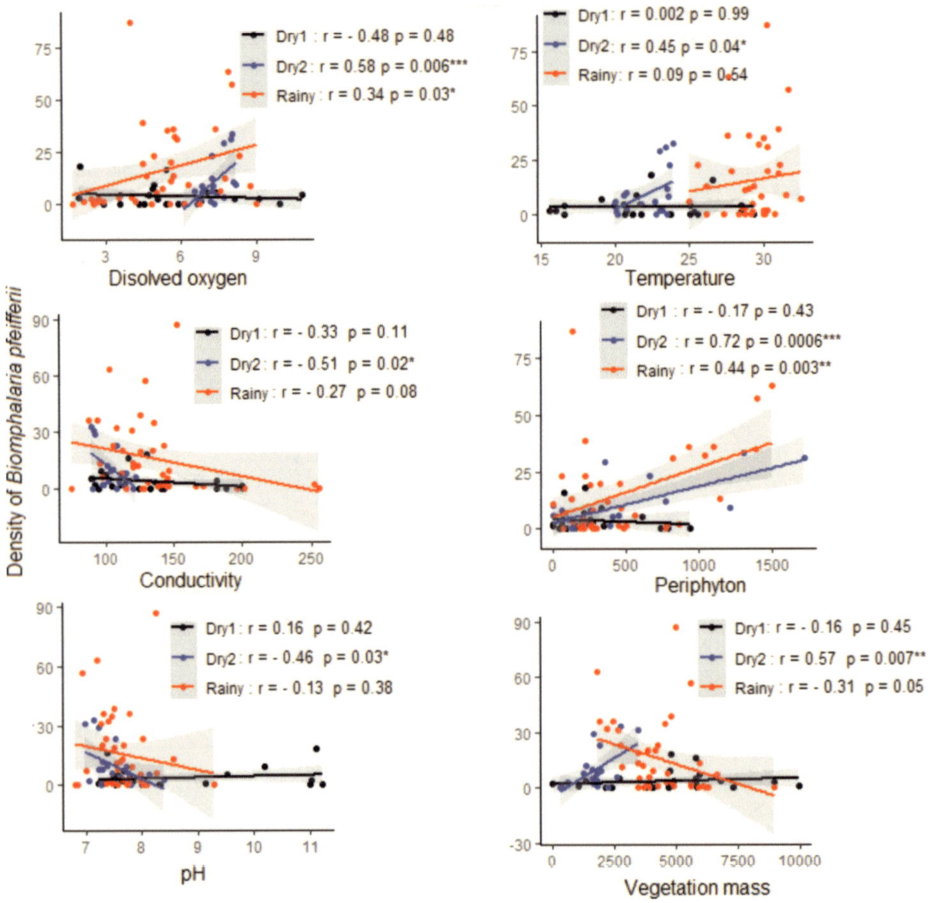

Figure 4.
Linear model assessing the significance of predictor-response relationships between environmental parameters and Biomphalaria pfeifferi *densities across the seasons.*

4. Discussion

4.1 Seasonal variations and environmental parameters driving the densities of snail hosts

Ecosystem changes in the delta of the Senegal River has created favorable biotopes to the development of intermediate host snails of human schistosomes, which may have included particular the physicochemical conditions favorable to their abundance [3]. Permanent freshwater is associated with aquatic floating vegetation such as *Ceratophyllum sp* and *Ludwigia sp* that acts as beneficial snails habitat in several areas in the Delta of Senegal River. The presence of the intermediate host *Biomphalaria pfeifferi* was previously described in several studies in the study area [3, 6–8]. In this area, *Biomphalaria pfeifferi* was the second most abundant intermediate host of human schistosomiasis after *Bulinus truncatus* [3]. Our study highlights that *B. pfeifferi* abundance is lowest during a Dry_1 season likely because of lower periphyton and water temperature than in the rainy season. Lower water temperature during Dry_1 season can limit snail growth and reproduction. For example, *Biomphalaria glabrata*, a neotropical species that is another intermediate host for *S. mansoni*, generally avoids thermal extremes and prefers temperatures from 27° to 32° C [9]. The greater density of *Biomphalaria* during periods of increasing temperature (19.9° to 32.5°C) during Dry_2 and Rainy is increasing abundance of periphyton that acts as snail food, as well as periods of higher dissolved oxygen, temperature and aquatic vegetation that acts as snail habitat. In contrast, during this same period, negative environmental influences on snail abundance including pH and conductivity are declining. Aquatic vegetation provides both habitat and food resources to snail hosts and is itself positively associated to the snail abundance [6], while high conductivity and pH impact negatively snails' intermediate hosts of human schistosome [3]. Although 55.1% of total vegetation mass was collected in the rainy season, vegetation was only a significant predictor of snails in the Dry2 season (10.9%). *Biomphalaria* species prefer areas of clear water on sandy and gravel bottoms, with stagnant water or with a very light current, with sometimes abundant aquatic vegetation [9]. Snail abundance was most associated with oxygen during the Dry_2 season, and *Biomphalaria pfeifferi* is known to be positively associated with dissolved oxygen [3], whereas pH in the range 5.0 to 7.5 is likely a weaker determinant of snails [9]. We suspect that snail density was not significantly correlated to water temperature because of every high temperature (32.5° C) we found during the rainy season. This driver could limit reproduction and other physiological functions or be lethal to the snail survival. *Biomphalaria pfeifferi* does best under warm stable conditions [10]. Thus, future studies should also consider non-linear relationships in their analyses. We found that the month of August was the most important for snail control because densities of host during the rainy season were fueled by ideal environmental conditions in freshwater (i.e: resource food availability that may increase host reproduction). During this period, rainwater transports organic wastes, fertilizer and pesticides from agricultural areas to adjacent waters. Such agrochemical runoff may affect the development aquatic vegetation and periphyton and pesticide pollution is a major driver in increasing the occurrence of host snails [11, 12]. Mesocosm studies support the assertion that fertilizer, herbicide, and insecticide, individually and as mixtures may be increasing the algae snails eat in the rainy season [13]. Moreover, such chemicals can decrease the densities of snail predators. Our findings may be context dependent to the Senegal River as seasonality may favor the dry season months in other contexts [14].

5. Conclusion

Given the widespread distribution of *Biomphalaria pfeifferi* in the Senegal River, understanding the seasonal variation and the principal drivers of snail abundance is important for local snails control to prevent upcoming human risk of *Schistosoma mansoni* infection. Our findings support that the rainy season is significantly associated ($p\text{-}value = 0.003$) with the abundance of intermediate hosts because of highly favorable environmental conditions for periphyton (snail food) and dissolved oxygen levels required by *B. pfiefferi*. Conditions were less ideal in the Dry_2 season than the rainy season, and, thus, more environmental parameters were significant for this period than in the rainy season. In contrast, environmental conditions were so poor in the Dry_1 season that few snails occurred and significant trends were absent. Our findings on the seasonal fluctuations of snail hosts are useful for targeting a snail control program during a time of year when it may be most effective to eliminate or reduce the vectors of schistosomiasis.

Author details

Sidy Bakhoum[1*], Christopher J.E. Haggerty[2], Cheikh Tidiane Ba[1], Nicolas Jouanard[3,4], Gilles Riveau[4,5] and Jason Robert Rohr[2]

1 Department of Animal Biology, Faculty of Sciences and Techniques, University Cheikh Anta Diop, Dakar, Senegal

2 Department of Biological Sciences, Environmental Change Initiative, Eck Institute of Global Health, University of Notre Dame, Notre Dame, IN, USA

3 Station d'Initiative Aquacole, Saint-Louis, Sénégal

4 Centre de Recherche Biomédical Espoir Pour La Santé, Saint-Louis, Sénégal

5 Center for Infection and Immunity of Lille (CIIL), Institut Pasteur de Lille, INSERM-U1019, CNRS UMR9017, Univ Lille, CHU Lille, Lille, France

*Address all correspondence to: bakhoumsidy09@gmail.com

References

[1] Altizer S, Dobson S, Hosseini P, Hudson P, Pascual M and Rohanil P. Seasonality and the dynamics of infectious diseases. Ecology Letters, 2006; *9*: 467-484

[2] WHO. The social context of schistosomiasis and its control (2008).

[3] Bakhoum S, Ndione RA, Haggerty CJE, Wolfe C, Sow S, Ba CT, Riveau G, Rohr Jason R. Influence des paramètres physico-chimiques sur la répartition spatiale des mollusques hôtes intermédiaires des schistosomes humains dans le delta du fleuve Sénégal. Med Sante Trop. 2019; 29: 61-67. doi: 10.1684/mst.2019.0883.

[4] Abdulkadir, F. M, Maikaje, D. B and Umar, Y. A. Ecology and Distribution of Freshwater Snails in Gimbawa Dam, Kaduna State, Nigeria. Nigerian Journal of Chemical Research. 2017; Vol. *22, No. 2*

[5] Moumane A, Karmaoui A., Karkouri J.A., Akchbab J. The Role of Community Participation in Fighting Schistosomiasis: Lessons from Akka Oases (Southern Morocco).2020; Chapter. DOI: 10.4018/978-1-7998-2197-7.ch006

[6] Haggerty CJE, Bakhoum S, Civitello DJ, De Leo GA, Jouanard N, Ndione RA, et al. Aquatic macrophytes and macro invertebrate predators affect densities of snail hosts and local production of schistosome cercariae that cause human schistosomiasis. PLoS Negl Trop Dis.2020; 14(7): e0008417. https://doi. org/10.1371/journal. pntd.0008417

[7] Ndir O. "Situation des Schistosomes au Sénégal". In: Chippaux JP. Lutte contre les schistosomes en Afrique de l'Ouest, 15-18 février 2000, Niamey-Cermes. Paris: éditions IRD, 292. p. 225-236.

[8] Diaw OT, Ndir O, Toupane MG. Guide de surveillance malacologique et de lutte contre les mollusques hôtes intermédiaires des bilharzioses. Ministère de la santé: Service national des grandes endémies, 1999.

[9] Grégoire, Y.Y., Mahama, T., Marcel, B.O., Emmanuel, T., Anne- Marie, B.O., & Alphonse, K.K. DYNAMIQUE DES POPULATIONS DE BIOMPHALARIA PFEIFFERI ET DE BULINUS GLOBOSUS EN ZONE D'ENDEMIE SCHISTOSOMIENNE EN COTE D'IVOIRE. European Scientific Journal, ESJ. *2014*; *10*.

[10] KN de, Wolmarans CT and Bornman M. Distribution and habitats of *Biomphalaria pfeifferi*, snail intermediate host of *Schistosoma mansoni*, in South Africa. Water SA. 2004; Vol. 30 No. 1

[11] Hoover Christopher M, Rumschlag Samantha L, Strgar Luke, Arakala Arathi, Gambhir Manoj, Leo Giulio A de, Sokolow Susanne H, Rohr Jason R, Remais Justin V, Effects of agrochemical pollution on schistosomiasis transmission: a systematic review and modelling analysis, The Lancet Planetary Health, Volume 4, Issue 7. 2020; Pages e280-e291, ISSN 2542-5196, https://doi.org/10.1016/S2542-5196(20)30105-4.

[12] Daniel J., Streicker Daniel G., Altizer Sonia (2017). Using host species traits to understand the consequences of resource provisioning for host–parasite interactions. Journal of Animal Ecology. J Anim Ecol. 2018; 87:511-525

[13] Halstead Neal T., Hoover Christopher M., Arakala, Arathi, Civitello David J., Leo Giulio A. De, Gambhir Manoj, Johnson Steve A., Jouanard Nicolas, Loerns Kristin A. McMahon Taegan A., Ndione Raphael A., Nguyen Karena, Raffe

Thomas R., Remais Justin V., Riveau Gilles, Skolow Susanne H. & Rohr Jason R. 2018 Agrochemicals increase risk of human schistosomiasis by supporting higher densities of intermediate hosts. NATURE COMMUNICATIONS. 2018; 9:837

[14] Rabone Muriel, Wiethase Joris Hendrik, Allan Fiona, Gouvras Anouk Nathalie, Pennance Tom, Hamidou Amina Amadou, Webster Bonnie Lee, Labbo Rabiou, Emery Aidan Mark, Garba Amadou Djirmay and Rollinson David, Freshwater snails of biomedical importance in the Niger River Valley: Evidence of temporal and spatial patterns in abundance, distribution and infection with Schistosoma spp. Parasites Vectors. *2019; 12:498*

Chapter 4

Impacts of Environmental Parameters on the Infectivity of Freshwater Snail

Wolyu Korma Erkano

Abstract

The successful transmission of the infective stage of the parasite (miracidia) depends on different factors. These free-living stages miracidia rely on their own stored energy and are directly exposed to environmental factors including disturbance resulting from pollution and human activities. There are different environmental factors that affect the cercarial infection of the snail. These include pH, temperature, salinity, dissolved oxygen, water hardness, habitat conditions, presence of predators and competitors, etc. Each of these factors may increase or decrease the freshwater snail's infectivity. The more hydrogen ion concentration in the aquatic habitat could have an effect on the maturation and physiology of the parasitic stage (miracidia), leading to impaired survival and reduced infectivity. In contrast, high temperature increases snail infectivity. While low dissolved oxygen in the aquatic environment results in low snail infectivity. Regarding the presence of predators can result in low snail infectivity by consuming the schistosome egg and the snails themselves. Total hardness also had a negative impact on the prevalence of snail infection. The hardness of the water results in the shell hardening of snails subsequently leads to low infection of snail by miracidia.

Keywords: environmental factors, freshwater snails, snail infectivity

1. Introduction

Snails belong to phylum – Mollusca and class – Gastropoda, which accounts for a large and highly diverse group of invertebrates. Many freshwater snails serve as an intermediate host for different digenetic trematodes that cause schistosomiasis, fascioliasis and other snail-borne diseases of humans and animals [1, 2]. Trematodes are a group of flatworms and utilize parasitism as a way of life. The development of larval digenean trematodes is a complex and multi-stage life cycle process that typically uses invertebrates as intermediate hosts and vertebrates as definitive hosts [3].

The eggs of the digenetic trematodes enter into the environment through feces or urine of the definitive host. Upon reaching to freshwater bodies, the eggs will hatch and release the first free-swimming larva of digenetic trematodes (miracidia) infecting a freshwater snail host. The larval developmental stages, such as sporocysts, rediae, and cercariae are completed in the freshwater snails. The infected snails shed thousands of the second-stage free-swimming larva called cercaria into the water. Then the cercaria enters into mammalian hosts through direct penetration of skin during contact with water bodies contaminated with human excreta containing parasite eggs. Adult parasites live in different sites of their definitive hosts, like the digestive system, circulatory system, respiratory, urinary, and reproductive systems [3–6]. The infection with digenetic trematodes cercaria gives rise to a disease for mammalian hosts. Understanding the trematodes lifecycle and transmission mechanism between freshwater vector snails and mammalian hosts is important to the control and elimination of the diseases [4, 6].

The miracidium is the first free-swimming larvae in a digenetic trematode's life cycle and is a main component in the transmission from vertebrates to molluscan hosts. However, different species of digenetic trematodes are known to have actively infecting free-swimming miracidia that quickly swim through water bodies searching out a suitable molluscan freshwater intermediate host [7].

Once the first free-swimming larvae hatch from the egg, different intrinsic and extrinsic parameters affect whether a successful freshwater snail penetration is achieved. Furthermore, eggs that take longer to embryonated may produce free-swimming larvae miracidia that are more sensitive to these environmental conditions [3].

The ability of free-swimming larvae miracidia to search its intermediate host is the primary stage in initiating a successful infection. Modification of different abiotic and biotic environmental conditions influences miracidia behavior [3, 8].

2. Environmental factors

Environmental condition change will lead to both direct (i.e.; physiological) and indirect (i.e.; interspecific interactions) influences on parasite transmission, some of which man increases disease while others will reduce infection [5]. In the transmission of digeneans, eggs and the free-living larvae that hatch from eggs, known as miracidia, play a crucial role. The eggs and miracidia are exposed to the environment, whether it is the host or the watery external habitat. To establish a suitable molluscan host, transmission stages must be able to detect a suitable molluscan host and ensure larval survival and infectivity. Actively infecting miracidia develop from eggs in the environment and swim actively to locate and penetrate the molluscan host (e.g., *Fasciolidae*, *Echinostomatidae*, *Philophthalmidae*, *Schistosomatidae*), whereas passively infecting miracidia hatch in the intestine of a suitable mollusk host (e.g., *Plagiorchiida*, *Hemiuroidea*, *Brachylaimoidea*) [9]. Miracidia hatch quickly upon adequate stimuli if eggs are fully embryonated (intrauterine development within adult parasite), or after a maturation period in the external environment if eggs are not fully embryonated (intrauterine development within adult parasite) [10].

The attachment to the freshwater snail surface and passage into tissue is thought to be the two steps of miracidium penetration into the host. This procedure is aided by the terebratorium, whose muscles aid penetration into the snail, and is likely accomplished in tandem with the histolytic secretion of the miracidia glands. Once located in the intermediate freshwater snail host, the miracidium has to be capable to penetrate the tissue [10].

Variations in environmental factors or physiological status could potentially account for some of the differences in infectivity in freshwater snail infectivity. Water quality and composition were shown to have a significant impact on snail infectivity [11]. Environmental factors both abiotic and biotic factors were the major determinant factor for snail infectivity [12].

The successful transmission of the infective stage of the parasite (miracidia) depends on different factors. These free-living stages miracidia rely on their own stored energy and are directly exposed to environmental factors including disturbance resulting from pollution and human activities. Among different environmental factors: temperature, pH, salinity, biological oxygen demand, dissolved oxygen, rainfall, hardness of water, predators and competitors are the most important determinants for snail infectivity [13].

2.1 Abiotic environmental factors

2.1.1 Temperature

Environmental factors frequently influence the outcomes of host–parasite interactions and, as a result, disease dynamics [14, 15]. Particularly as ambient temperature has a significant impact on these dynamics [16, 17]. Understanding such effects is becoming increasingly relevant as a result of anthropogenic climate change, as climate change leads to an increase in average air temperature and more frequent occurrence of extreme weather events, such as summer heatwaves [18, 19]. High temperatures are thought to affect host–parasite interactions by lowering host resistance to infections as a result of a poor immunological function [20, 21]. This is because the immune system is commonly thought to be the primary physiological barrier against parasites [22]. It can be difficult to predict the effects of temperature on host–parasite interactions based just on the information of host immune defense. This is because parasite infection methods, as well as host traits other than immune function, can play a role in how interactions turn out [23]. Temperature, for example, can have a direct impact on parasite infective stages (e.g., survival l, movement), altering infection success [7, 24]. Furthermore, temperature influences the metabolism of ectothermic species and may alter their physiological characteristics [25].

In the two experimental laboratory setup, freshwater snails kept at 25°C had higher parasite infection success than snails kept at 15°C over the long term. In the short-term treatment, maintaining a temperature of 25°C prior to parasite contact resulted in fewer infections than maintaining a temperature of 15°C. The experimental snails were assigned randomly in to one of the two temperature (15°C and 25°C) [23]. The other study also shows when the average temperature was 30.5°C, the average relative humidity was 73 percent, rainfall was 145.7 mm, and pan evaporation was 4.0 mm, the highest infection occurred. The infection of snails and meteorological parameters (temperature, relative humidity, and rainfall) were found to have a positive relationship [26].

At low and high temperatures, the rate of infectivity drop is the fastest. Elevated temperatures had a deleterious impact on hatching success and infection rates, corroborating our results of morphological degradation and motility patterns (**Figure 1**). Because the lower hatching and infection rates associated with higher temperatures suggest that temperature has a direct impact on trematode biological integrity during its "free-living" stages, and ultimately transmission success, independent of temperature-mediated snail susceptibility [27].

While higher temperatures were linked to higher infection rates, these findings reflect the influence of temperature on ectothermic snails, including their feeding behavior, rather than the long-term effects of temperature on parasite physiology. Environmental conditions alter miracidia and their snail hosts, creating complexity through nonlinear temperature-dependent processes and time-dependent metabolic trade-offs that favor particular and restricted transmission windows. This

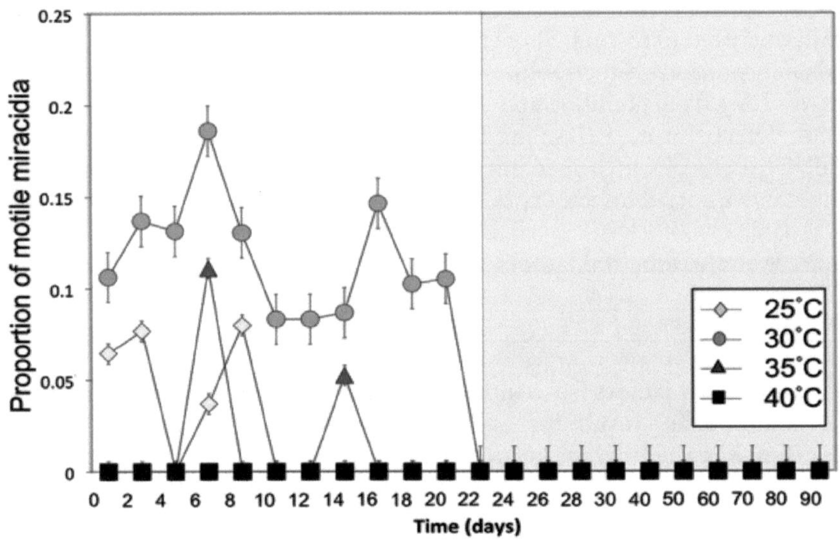

Figure 1.
Proportion of motile miracidia over time and across temperatures [27].

experiment is done in randomly distributed snails in five replication per temperature treatment (**Figure 2**) [27].

Although comparative investigations of different species have so far been limited to cercariae, trematodes in particular appear to have a complicated and variable relationship with temperature [28, 29]. High snail infection was obtained in the temperature ranges between 22.16°C to 24.66°C. Temperature is one of the most important parameters for snail and trematode larval growth [30], hence these temperature ranges are ideal for trematode larva infection in host snails [5].

With respect to different larval stages (sporocyst, redia, and cercaria) of the trematode in freshwater snail host, direct involvement of metabolic energy usage

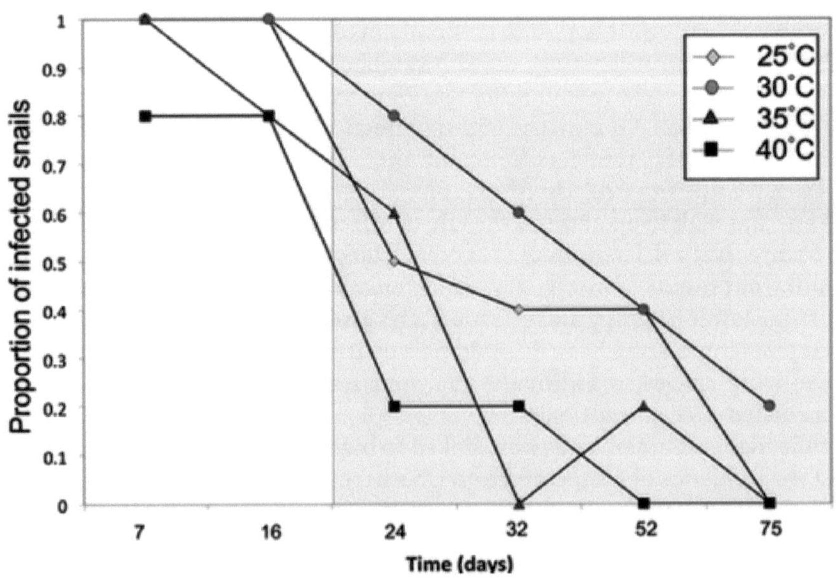

Figure 2.
Proportion of snails infected over time and across temperatures [27].

(i.e., oxygen consumption rate), as well as abiotic environmental parameters such as temperature, pH, oxygen, and free carbon dioxide, is required [5].

Because of its impact on the parasite and its host, the effect of temperature on vector-borne diseases such as schistosomiasis is well-known. Snail hosts' ability to survive and reproduce is temperature sensitive and varies between species [31]. The temperature has an effect on parasite viability, as evidenced by the reduced infectivity of cercariae emerging from sporocysts in snails kept at temperatures of 23–25°C. As the snail maintenance temperatures are dropped, the pre-patent period tends to be lengthened [32].

Temperature is one of the most important environmental elements governing miracidia life span, and it can have a big impact on transmission viability. It has a direct impact on activity, with larger amounts of movement occurring when the temperature rises. Understanding the thermal biology of trematodes is crucial for predicting parasite population dynamics in the face of climate change. The effects of the climate on both hosts and parasites can alter the extent and intensity of parasitism, with the temperature being a particularly important role [7, 33].

2.1.2 Rainfall

The impact of rainfall patterns on the density of intermediate snail populations and infection rates, stressing how the presence and seasonal vigor of breeding sites and foci dictate the periodicity of schistosomiasis transmission in a specific area. Snail infection rates were high until the dry season began when population density began to fall. The initial captures following rains delivered adult snails that were already infected, demonstrating that they may maintain *S. mansoni* infection during aestivation and are capable of shedding cercariae and transmitting the disease as long as weather conditions permit life in the breeding site [34].

According to the studies, there is no infected freshwater snails found between November 2013 and March 2014 except December; but staring in April, infected snails were found coinciding with the rainy season (**Figure 3**) [35].

2.1.3 Salinity and pH

High electrical conductivity and chloride in water bodies indicated that the total dissolved salts in the water were high [36]. The vulnerability of miracidium to salinity is of course strongly related to whether a parasite inhabits a freshwater habitat. The miracidium can withstand slight salt concentrations without affecting their survival [13]. The freshwater that has high salinity may lead to a reduction of miracidium infectivity, subsequently, it results in low infection in snails [37].

Regarding the pH, it had a negative association with freshwater snail infection. The more hydrogen ion concentration in the aquatic habitat could have an effect on the maturation and physiology of the first free-swimming stage (miracidia), leading to impaired survival and reduced infectivity [5, 13].

The study shows freshwater snails and the parasites are well-adapted to high-salinity and alkaline coastal habitats (**Figures 4** and **5**) [34]. The trematode life cycle, including embryonic development, is influenced by environmental conditions like pH. The impact of pH on trematode larvae hatchability in aquatic settings is important. Environmental factors influence distinct stages of parasitic fluke life cycles and, as a result, their epidemiology [38, 39].

Different researchers have found that the maximum hatching rate of miracidia occurs at a neutral pH level [38]. At a pH of 5–10, the hatchability of eggs is

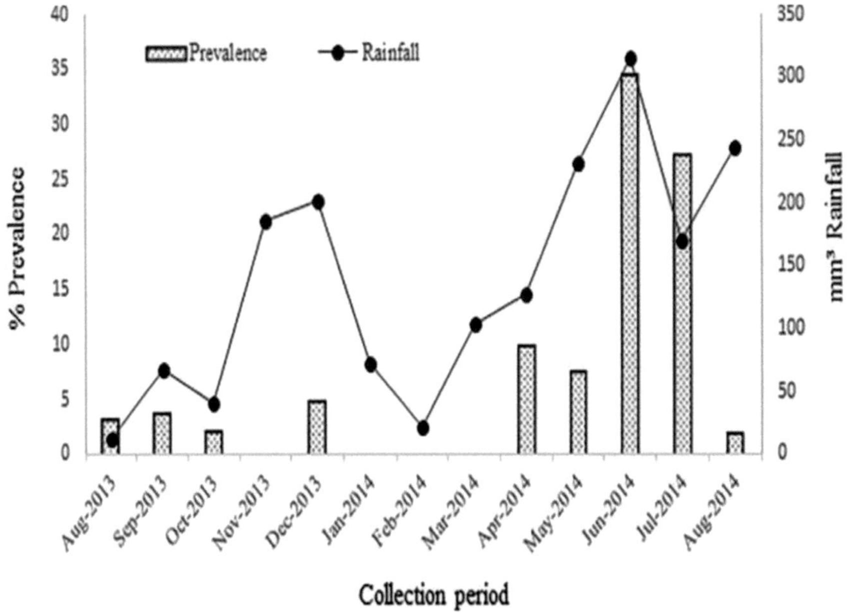

Figure 3.
Correlation between snail infection and rainfall [35].

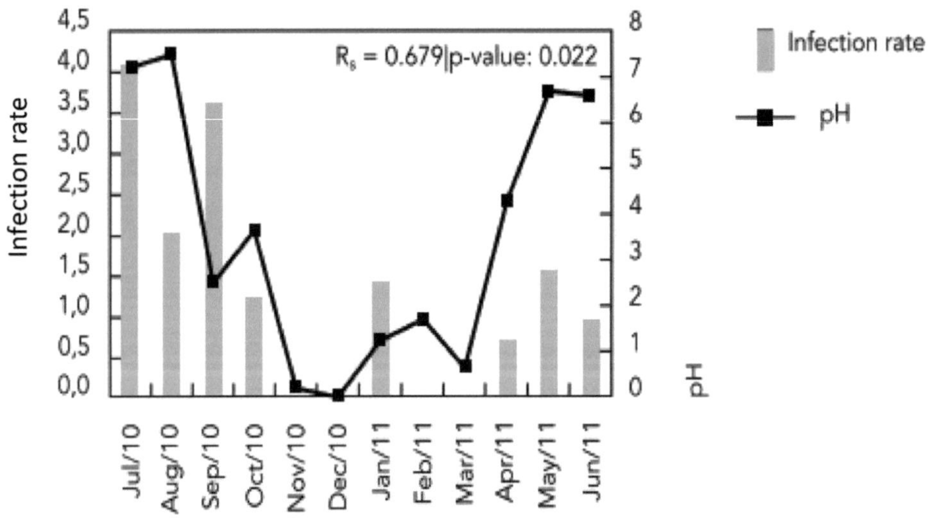

Figure 4.
Relationship between snail infection and pH [34].

decreased. The proliferation of miracidia in trematode eggs was impacted by pH variations. Subsequently, it minimizes the infection rate of intermediate snail hosts. Several studies have demonstrated that higher and lower pH levels affect the time it takes for trematodes to hatch and the number of sterile eggs they produce. [38, 39].

The influence of water pH on the toxicity of heavy metals can have an indirect impact on the life cycle and multiplication of parasitic trematodes [40]. pH has an impact on the life cycle of trematodes as well as their intermediate hosts, such as snails, from an epidemiological perspective. Freshwater snail distribution is influenced by water quality, notably pH [41]. The pH range of 7.2–7.5 is ideal for snail activity and population propagation [42]. Snails living in freshwater may be

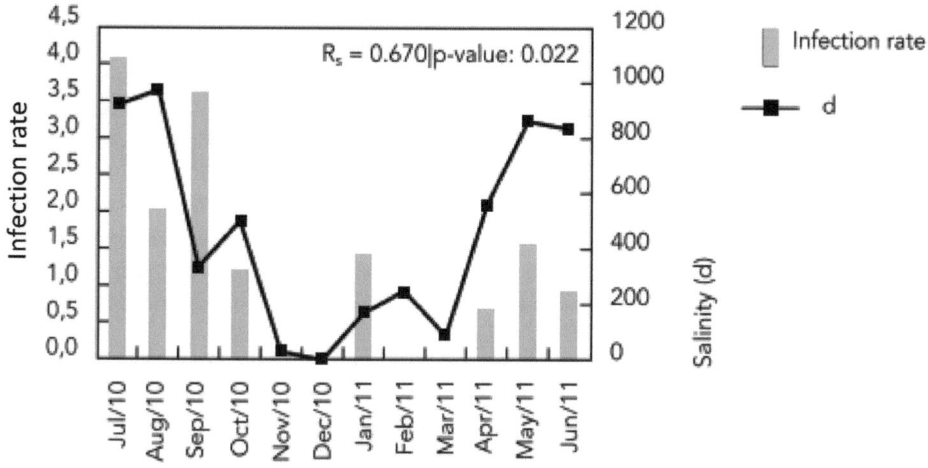

Figure 5.
Relationship between snail infection and salinity [34].

killed by low pH levels. This reduces freshwater snail abundance and subsequently results in low snail infection. This is due to the high abundance of freshwater snails that could promote cercaria transmission as the access distance for the miracidia is reduced as a result more snails get infected, which could be because why the area with lowest freshwater snails abundance had the lowest infection in snail [37]. The alkaline pH of about eight is ideal for the multiplication of freshwater snails [41].

2.1.4 Dissolved oxygen and BOD_5

The BOD5 had a positive relationship with the freshwater snail infection. These might be due to the presence of organic pollution which increases BOD5 and subsequently increases the infection of freshwater snails. Whereas dissolved oxygen concentration had a negative relationship with snail infectivity. The low dissolved oxygen was an indication of the presence of organic pollution in the aquatic environment [43]. This organic pollution is beneficial and expands the habitat of freshwater snail hosts and subsequently, more snails might be present. And also, it might have a trematode egg in the waste. With the high availability of freshwater snail hosts and trematode eggs, the probability of a miracidium searching and infecting a snail is higher. As a result, the cercarial infection in the freshwater snail host could be increased [44].

2.1.5 Hardness of water

Freshwater snails need calcium for shell hardening, egg production and are unable to get it in the water with less than 2 mg/l calcium. The ratio of calcium to magnesium is significant for calcium uptake by the snails. The freshwater snail that lives in the water with a low concentration of calcium was highly infected than in high concentration calcium [45].

2.2 Biotic environmental factors

2.2.1 Predators

The prevalence of freshwater snail infection is also affected by biological factors. The predators including Belostomatidae, Naucoridae, Nepidae, Gomphidae, and

Gyrinidae were prey on the freshwater snail and consume trematode larva miracidia [46, 47]. As a result, the prevalence of freshwater snail infection with trematode might be reduced. And also, the snail infectivity might be indirectly influenced by competitors like Thiaridae, Physidae, and Lymnaeidae [46, 48]. This results in an abundance of infected snails were decreased in the presence of predators and competitors [49].

Furthermore, biotic factors such as toxins produced by hosts, non-hosts, predators, or decoy organisms may act in concert with abiotic factors to expose free-living endohelminth stages to a diverse range of hazards on their way to the next host, further modulating transmission dynamics and infection patterns in a host population [27, 50].

The functioning of aquatic relationships is influenced by predatory and competitive forces. Predators often cause trophic cascades in which resource dynamics and energy balances in the environment are affected [51]. Such trophic cascades could be detected in Zambia [52] and Malawi [53] as a general reduction in aquatic biodiversity due to strong fishing pressure. Overfishing has reduced the predation pressure on snails, and fish mortality from metals, pesticides, and other chemical pollutants has reduced the top-down regulation of trophic interactions [32]. These all conditions reduce the abundance of snails in the aquatic environment and subsequently results in low snail infection. This is due to the high abundance of freshwater snails that could promote cercaria transmission as the access distance for the miracidia is reduced as a result more snails get infected, which could be because why the area with the lowest freshwater snails abundance had a low infection in snail [37].

2.2.2 Competitors

The non-host snail had the most significant influence on miracidia, causing them to enhance their host-finding behavior and penetration attempts. They discovered that failed penetration attempts resulted in severe mage and tiredness, making infecting the host snail more difficult. In addition, failed attempts caused them to shed penetration glands, making infection more difficult. When the media was infected with sew, the rate of miracidial infection in host snails decreased. This could be because of the fact that the sew contained a harmful material discharged by the none host snails, triggering a host snails regulatory response [3].

Snail introductions produce comparable outcomes, as they can infest wide regions of new suitable habitat and establish dominance. These have both bad and positive consequences since they cause economic damage and disruption to the biodiversity of the area in question, as well as positive results as a result of their competitive and predatory behaviors against hosts of dangerous parasites like schistosomes [32, 54].

2.3 Habitat conditions

Some habitat conditions like silt, organic, chlorophyll-a, canopy cover, riparian vegetation and freshwater snail abundance have an association with snail infectivity. These factors may be contributing to the development of the hotspot include the vulnerability of snails to infection, probability of interaction with viable trematode eggs, and suitability of snail habitat [55]. This study shows the habitat with silt, organic, shadow, and muddy grass-grown highly favored by the snails and the majority of infected snails were found in silt substrates, and greater than 50% riparian vegetation. The high abundance of snails that could promote cercaria transmission as the access distance for the miracidia is reduced as a result more snails get infected, which could be because of why the site with lowest snail abundance had a low infection in snail [37].

3. Snail borne diseases

There are different intermediate host snail species in the world freshwater bodies which cause snail borne parasitic diseases like schistosomiasis, paragonimiasis, fascioliasis, fasciolopsiasis, angiostrongyliasis, clonorchiasis, and opisthorchiasis. These diseases are the most important parasitic disease which remains crucial to public health issues worldwide, mainly in developing countries. Millions of people in 90 countries have suffered from snail borne disease, in which snails are intermediate hosts and transmitting vectors. These diseases also resulting in extensive socioeconomic burdens in many tropical and sub-tropical countries [1].

Specifically, human schistosomiasis is one of the most prevalent parasitic infections in the world and found in 52 countries. A report from WHO indicated that 219.9 million people worldwide are estimated to be affected by schistosomiasis, of which it is estimated that at least 90.4% of those requiring treatment for schistosomiasis live in Africa. This disease caused a loss of 2.5 million disability-adjusted life years (DALYs) [6]. It is the second most widespread parasitic disease after malaria and killing an estimated 300,000 people each year in the African region alone and 163 million population need treatment in sub-Saharan Africa [56, 57].

Schistosomiasis is caused by six species of trematodes from the genus *Schistosoma*: *Schistosoma mansoni*, *S. haematobium*, *S. japonicum*, *S. intercalatum*, *S. guineensis* and *S. mekongi*. The predominant causes of disease are *S. mansoni* and *S. haematobium* in tropical and subtropical regions, particularly in sub-Saharan Africa [6]. *Schistosoma mansoni* and *S. haematobium* trematodes are transmitted by two primary snail species of the genus *Biomphalaria* and *Bulinus* respectively, which are widely distributed throughout African countries [58].

Environmental and endogenous variables influence the development of *Schistosoma mansoni*'s life cycle in its intermediate host [58]. Mollusk infections occur in freshwater bodies contaminated by schistosome-infected people's feces. The existence of the Biomphalaria mollusk, as well as lack of or inadequate sanitation, human cultural habits, and the parasite's life cycle, all contribute to the parasite's persistence and, as a result, the disease's geographic spread [59–61].

The complete spectrum of direct and indirect effects on host and parasite life histories determines the impact of abiotic environmental factors on infectious disease dynamics. Importantly, these effects will go beyond a simple shift in host or parasite geographic distribution to involve a considerable modification in the physiological and temporal interaction between host and parasites, thereby altering disease dynamics in natural populations [62].

4. Conclusion

Environmental changes along the borders of water bodies aimed at diminishing snail habitats are one method of snail control. In both snail and parasite intervention during control operations, knowledge of the environmental factors in snail infectivity in endemic areas is vital. It will be feasible to take efforts toward preventing and regulating the harmful impacts of parasitic trematode outbreaks by knowing the ecological drivers of their infection, growth and propagation. As a result, it may cause overt harm to humans and livestock, as well as economic losses in the future. In man-made systems, where the design phase includes proper technical activities, environmental change may be optimal. Control techniques must be tailored to the ecology of the host snails as well as the social characteristics of the affected group, and executed on an individual basis.

Conflict of interest

The author declares no conflict of interest.

Author details

Wolyu Korma Erkano
School of Public Health, Woldia University, Woldia, Ethiopia

*Address all correspondence to: wolyukorma53@gmail.com

References

[1] Lu, X.-T., Q.-Y. Gu, Y. Limpanont, L.-G. Song, et al. Snail-borne parasitic diseases: An update on global epidemiological distribution, transmission interruption and control methods. Infectious diseases of poverty. 2018; 7:1: 28.

[2] Tigga, M., R. Bauri, A. Deb, and S. Kullu. Prevalence of snail's intermediate host infected with different trematodes cercariae in and around Ranchi. Veterinary World. 2014; 7:8: 630-634.

[3] Rolfsen, B., Compatibility factors of Fascioloides magna miracidia and four sympatric snail species: Miracidial behavior and snail response. 2015, Eastern Illinois University.

[4] Colley, D.G., A.L. Bustinduy, W.E. Secor, and C.H. King. Human schistosomiasis. The Lancet. 2014; 383:9936: 2253-2264.

[5] Sunita, K., P. Kumar, and D. Singh. Abiotic environmental factors and infection of Fasciola gigantica in vector snail Lymnaea acuminata. Researcher. 2012; 4:8: 49-53.

[6] WHO, Schistosomiasis and soiltransmitted helminthiases: numbers of people treated in 2017. 2018.

[7] Morley, N. Thermodynamics of miracidial survival and metabolism. Parasitology. 2012; 139:12: 1640-1651.

[8] Munoz-Antoli, C., M. Trelis, M. Gozalbo, R. Toledo, et al. Interactions related to non-host snails in the host-finding process of Euparyphium albuferensis and Echinostoma friedi (Trematoda: Echinostomatidae) miracidia. Parasitology Research. 2003; 91:5: 353-356.

[9] Galaktionov, K.V. and A. Dobrovolskij, The biology and evolution of trematodes: an essay on the biology, morphology, life cycles, transmissions, and evolution of digenetic trematodes. 2013: Springer Science and Business Media.

[10] Born-Torrijos, A., A.S. Holzer, J.A. Raga, G.S. van Beest, et al. Description of embryonic development and ultrastructure in miracidia of Cardiocephaloides longicollis (Digenea, Strigeidae) in relation to active host finding strategy in a marine environment. Journal of morphology. 2017; 278:8: 1137-1148.

[11] Prugnolle, F., T. De Meeûs, J. Pointier, P. Durand, et al. Geographical variations in infectivity and susceptibility in the host-parasite system Schistosoma mansoni/ Biomphalaria glabrata: no evidence for local adaptation. Parasitology. 2006; 133:3: 313-319.

[12] Mardin, S., T.W. Sardjono, L.E. Fitri, and A. Ramadhan. Effect of ecological factors on snails infection by Schistosoma japonicum in Central Sulawesi, Indonesia. American Journal of Environmental Sciences. 2018; 14:2: 55-62.

[13] Pietrock, M. and D.J. Marcogliese. Free-living endohelminth stages: at the mercy of environmental conditions. Trends in parasitology. 2003; 19:7: 293-299.

[14] Altizer, S., A. Dobson, P. Hosseini, P. Hudson, et al. Seasonality and the dynamics of infectious diseases. Ecology letters. 2006; 9:4: 467-484.

[15] Harvell, C.D., C.E. Mitchell, J.R. Ward, S. Altizer, et al. climate warming and disease risks for terrestrial and marine biota. Science. 2002; 296:5576: 2158-2162.

[16] Macnab, V. and I. Barber. Some (worms) like it hot: Fish parasites grow

faster in warmer water, and alter host thermal preferences. Global Change Biology. 2012; 18:5: 1540-1548.

[17] Poulin, R. and K. Mouritsen. Climate change, parasitism and the structure of intertidal ecosystems. Journal of helminthology. 2006; 80:2: 183-191.

[18] Diffenbaugh, N.S., J.S. pal, R.J. Trapp, and F. Giorgi. Fine-scale processes regulate the response of extreme events to global climate change. Proceedings of the National Academy of Sciences. 2005; 102:44: 15774-15778.

[19] Meehl, G.A. And C. Tebaldi. More intense, more frequent, and longer lasting heat waves in the 21st century. Science. 2004; 305:5686: 994-997.

[20] Murdock, C., K.P. Paaijmans, A.S. Bell, J.G. King, et al. Complex effects of temperature on mosquito immune function. Proceedings of the Royal Society B: Biological sciences. 2012; 279:1741: 3357-3366.

[21] Roth, O., J. Kurtz, and T.B. Reusch. A summer heat wave decreases the immunocompetence of the mesograzer, Idotea baltica. Marine Biology. 2010; 157:7: 1605-1611.

[22] Janeway, C.A. Immunobiology: The immune system in health and disease New York/London. Garland/Current Biology. 2005; 9: 13-14.

[23] Leicht, K. And O. Seppälä. Infection success of Echinoparyphium aconiatum (Trematoda) in its snail host under high temperature: Role of host resistance. Parasites vectors. 2014; 7:1: 1-6.

[24] Fried, B. and E. Ponder. Effects of temperature on survival, infectivity and in vitro encystment of the cercariae of Echinostoma caproni. Journal of Helminthology. 2003; 77:3: 235-238.

[25] Hofmann, G.E. and A.E. Todgham. Living in the now: Physiological mechanisms to tolerate a rapidly changing environment. Annual review of physiology. 2010; 72: 127-145.

[26] Niaz, S., T. Akhtar, A. Hasanat, and A. Qureshi. Prevalence of snails and Schistosome cercariae and correlation with meteorological factors in Punjab, Pakistan. Iranian Journal of Veterinary Research. 2013; 14:2: 161-164.

[27] Echaubard, P., T. León, K. Suwanatrai, J. Chaiyos, et al. Experimental and modelling investigations of Opisthorchis viverrini miracidia transmission over time and across temperatures: Implications for control. International journal for parasitology. 2017; 47:5: 257-270.

[28] Koprivnikar, J. And R.J.J.o.P. Poulin. Interspecific and intraspecific variation in cercariae release. 2009; 95:1: 14-19.

[29] Morley, N. Thermodynamics of cercarial survival and metabolism in a changing climate. Parasitology. 2011; 138:11: 1442-1452.

[30] Njoku-Tony, R. Effect of some physico-chemical parameters on abundance of intermediate snails of animal trematodes in Imo state, Nigeria. Researcher. 2011; 3:4: 15-21.

[31] Appleton, C., H. Madsen, and Management. Human schistosomiasis in wetlands in southern Africa. Wetlands Ecology. 2012; 20:3: 253-269.

[32] Monde, C., S. Syampungani, and P. Van den Brink. Exploring the potential of host-environment relationships in the control of schistosomiasis in Africa. African Journal of Aquatic Science. 2015; 40:1: 47-55.

[33] Mas-Coma, S., M.A. Valero, and M.D. Bargues. Climate change effects on trematodiases, with emphasis on zoonotic fascioliasis and

schistosomiasis. Veterinary parasitology. 2009; 163:4: 264-280.

[34] Leal Neto, O.B., E.C.d.S. Gomes, F.J.M.d. Oliveira Junior, R. Andrade, et al. Biological and environmental factors associated with risk of schistosomiasis mansoni transmission in Porto de Galinhas, Pernambuco State, Brazil. Cadernos de saude publica. 2013; 29: 357-367.

[35] Calasans, T.A.S., G.T.R. Souza, C.M. Melo, R.R. Madi, et al. Socioenvironmental factors associated with Schistosoma mansoni infection and intermediate hosts in an urban area of northeastern Brazil. Plos one. 2018; 13:5: e0195519.

[36] Kazibwe, F., B. Makanga, C. Rubaire-Akiiki, J. Ouma, et al. Ecology of Biomphalaria (Gastropoda: Planorbidae) in Lake Albert, Western Uganda: Snail distributions, infection with schistosomes and temporal associations with environmental dynamics. Hydrobiologia. 2006; 568:1: 433-444.

[37] Rowel, C., B. Fred, M. Betson, J.C. Sousa-Figueiredo, et al. Environmental epidemiology of intestinal schistosomiasis in Uganda: Population dynamics of Biomphalaria (Gastropoda: Planorbidae) in Lake Albert and Lake Victoria with observations on natural infections with digenetic trematodes. BioMed research international. 2015; 2015: 11.

[38] M Al-jibouri, M., H. R Hassan, and S. H Al-Mayah. Ecological factors affecting on eggs development and life span of meracidia of *Fasciola gigantica*. Karbala journal of pharmaceutical sciences. 2010; 1:1: 74-80.

[39] Yakhchali, M. and K. Bahramnejad. Inhibition effect of pH on the hatchability of Fasciola miracidia under laboratory conditions. Iranian journal of parasitology. 2016; 11:1: 30.

[40] Islam, M.N., G.R. Port, and A.J. McLachlan. The biology of Lymnaea peregra (Muller) (Gastropoda: Pulmonata: Basommatophora) with special reference to the effects of herbicides on its reproduction. Biological science 2001.

[41] De Francesco, C.G. and F.I. Isla. Distribution and abundance of hydrobiid snails in a mixed estuary and a coastal lagoon, Argentina. Estuaries. 2003; 26:3: 790-797.

[42] Karimi, G., M. Derakhshanfar, and H. PEYKARI. Population density, trematodal infection and ecology of Lymnaea snails in Shadegan, Iran. Archives of razi institute 2004.

[43] De Troyer, N., S.T. Mereta, P.L. Goethals, and P. Boets. Water quality assessment of streams and wetlands in a fast growing east African city. Water. 2016; 8:4: 123.

[44] Grimes, J.E., D. Croll, W.E. Harrison, J. Utzinger, et al. the roles of water, sanitation and hygiene in reducing schistosomiasis: A review. Parasites and vectors. 2015; 8:1: 1-16.

[45] Mostafa, O.M.S. Effects of Schistosoma mansoni and Schistosoma haematobium infections on calcium content in their intermediate hosts. Parasitology research. 2007; 101:4: 963-966.

[46] El Bardicy, S., M. Tadros, F. Yousif, and S. Hafez. Predatory activity of *Psychoda alternata* Say (Diptera: Psychodidae) larvae on *Biomphalaria glabrata* and Lymnaea natalensis snails and the free-living larval stages of *Schistosoma mansoni*. Australian Bas Appl Sci J. 2009; 3: 4503-4509.

[47] Yousif, F., S. Hafez, S. El Bardicy, M. Tadros, et al. Experimental evaluation of Candonocypris novaezelandiae (Crustacea: Ostracoda) in the biocontrol of Schistosomiasis mansoni

transmission. Asian Pacific journal of tropical biomedicine. 2013; 3:4: 267-272.

[48] Armúa de Reyes, C. and A.L. Estévez. Predation on Biomphalaria sp. (Mollusca: Planorbidae) by three species of the genus Belostoma (Heteroptera: Belostomatidae). Brazilian Journal of Biology. 2006; 66:4: 1033-1035.

[49] Swartz, S.J., G.A. De Leo, C.L. Wood, and S.H. Sokolow. Infection with schistosome parasites in snails leads to increased predation by prawns: Implications for human schistosomiasis control. Journal of Experimental Biology. 2015; 218:24: 3962-3967.

[50] Thieltges, D., K. Jensen, and R. Poulin. The role of biotic factors in the transmission of free-living endohelminth stages. Parasitology. 2008; 135:4: 407-426.

[51] S. Greig, H. and A. R. McIntosh. Indirect effects of predatory trout on organic matter processing in detritus-based stream food webs. Oikos. 2006; 112:1: 31-40.

[52] Musumali, M.M., S. Heck, S.M. Husken, and M. Wishart. Fisheries in Zambia: An undervalued contributor to poverty reduction. The World fish centre/World bank policy. 2009.

[53] Stauffer JR, M.H., Schistosomiasis in Lake Malawi and the potential use of indigenous fish for biological control. 2012, Rijeka: InTech. 119-140.

[54] Pointier, J.-P., R. DeJong, L.T. Tchuenté, T. Kristensen, et al. A neotropical snail host of Schistosoma mansoni introduced into Africa and consequences for the schistosomiasis transmission: Biomphalaria tenagophila in Kinshasa (Democratic Republic of Congo). Acta Tropica. 2005; 93:2: 191-199.

[55] Tolley-Jordan, L.R. and J.M. Owen. Habitat influences snail community structure and trematode infection levels in a spring-fed river, Texas, USA. Hydrobiologia. 2008; 600:1: 29-40.

[56] Lai, Y.-S., P. Biedermann, U.F. Ekpo, A. Garba, et al. Spatial distribution of schistosomiasis and treatment needs in sub-Saharan Africa: A systematic review and geostatistical analysis. The Lancet infectious diseases. 2015; 15:8: 927-940.

[57] Olveda, D.U., Y. Li, R.M. Olveda, A.K. Lam, et al. Bilharzia: Pathology, diagnosis, management and control. Tropical medicine and surgery. 2013; 1:4.

[58] Coelho, J.R. and F.S. Bezerra. The effects of temperature change on the infection rate of Biomphalaria glabrata with Schistosoma mansoni. Memórias do Instituto Oswaldo Cruz. 2006; 101:2: 223-224.

[59] Rocha, T.J.M., M.C.S. Santos, M.V.M.d. Lima, C.M.L. Calheiros, et al. Epidemiological aspects and distribution of infection cases by *Schistosoma mansoni* in municipalities in the Alagoas State, Brazil/Aspectos epidemiológicos e distribuição dos casos de infecção pelo *Schistosoma mansoni* em municípios do Estado de Alagoas, Brasil. Rev Pan-Amaz Saude. 2016: 27-32.

[60] Scholte, R.G., L. Gosoniu, J.B. Malone, F. Chammartin, et al. Predictive risk mapping of schistosomiasis in Brazil using Bayesian geostatistical models. Acta tropica. 2014; 132: 57-63.

[61] Teles, H.M.S. Geographic distribution of Schistosoma mansoni transmitter snail species in state of São Paulo. Revista da Sociedade Brasileira de Medicina Tropical. 2005; 38:5: 426-432.

[62] Johnson, P.T. and S.H. Paull. The ecology and emergence of diseases in fresh waters. Freshwater Biology. 2011; 56:4: 638-657.

Chapter 5

Recent Trends in Freshwater Pearl Farming in India

Shailesh Saurabh, Sweta Pradhan and Sonal Suman

Abstract

Cultured pearls have an important place in international trade. The Vedas, the Bible, and the Koran all mentioned pearls, and they are regarded as one of the highest honours. Pearls are generated in nature when an irritant, such as a sand grain or a parasite, is swept into the pearl molluscs and lodged within it, where it is coated with micro-layers of nacre, a lustrous substance made up of 80–90 per cent aragonite crystals of $CaCO_3$. The ICAR-Central Institute of Freshwater Aquaculture (CIFA), Kausalyaganga, Bhubaneswar, India, has created a base technology for cultivating pearls in freshwater habitats, recognising the scope and value of freshwater pearl production. Indian pond mussel, *Lamellidens marginalis* is the major species used in freshwater pearl aquaculture. In addition, ICAR-CIFA has pioneered a novel feature of freshwater pearl farming. The Institute has also taken the lead in disseminating freshwater pearl culture technology to the country's fish farming communities, entrepreneurs, researchers, and students to build a sustainable model for the country's socio-economic development. In this chapter, we will briefly cover pearls and their types, their historical significance, the spread of pearl mussels of freshwater origin in various countries, pearl biomineralisation, pearl farming techniques, and factors affecting pearl quality, among other things.

Keywords: cultured pearl, *Lamellidens marginalis*, CIFA, mussels, farming

1. Introduction

The "Queen of Gems," pearl, has a long history of cultural significance and great commercial demand, making it one of the most profitable aquaculture ventures in countries with extensive bivalve resources. Pearl harvesting was once restricted to wild aquatic resources, but with the innovation and standardisation of pearl producing processes, it has become a profitable aquaculture practice. The stimulation of a defence mechanism, where some kind of irritant prompts the bivalves to exude the glossy nacre, which coats the irritant by progressively evolving into layers to create the pearl, is the essential biology responsible for the production of pearly gems [1]. The pearl production process and cultured pearls are driven by this defence event, which is replicated in captivity. However, the ability to make the gem is not found in all molluscan bivalves; rather, only those species with a nacreous layer beneath their shell may make pearls [2].

Commercial pearl production is currently taking place in a number of nations throughout the world, including China, Japan, Australia, Indonesia, French Polynesia, Cook Islands, Philippines, India, Sri Lanka, Bangladesh, Myanmar, Thailand, Malaysia, and Mexico [3, 4]. China and Japan are the main producers of

IntechOpen

freshwater and marine pearls, respectively, whereas China is the world's biggest producer of pearls, including both marine and freshwater pearls, with 3540 tonnes produced, accounting for 98% of global pearl production [2]. According to the recent FAO (2020) data, freshwater pearl production is continuing to decline, both in terms of volume and value [5], as seen in **Figures 1** and **2**. It's difficult to answer the question of what can be causing the drop in freshwater pearl output year after year. However, a number of causes, including urbanisation, pollution, climate change, longer pearl production cycles, habitat degradation, sedimentation, a scarcity of qualified manpower, pesticide issues, and the lack of implantation kits, may be to blame for the fall in freshwater pearl output. It is imperative that actions be taken by fisheries departments, policymakers, planners, research organisations, financing

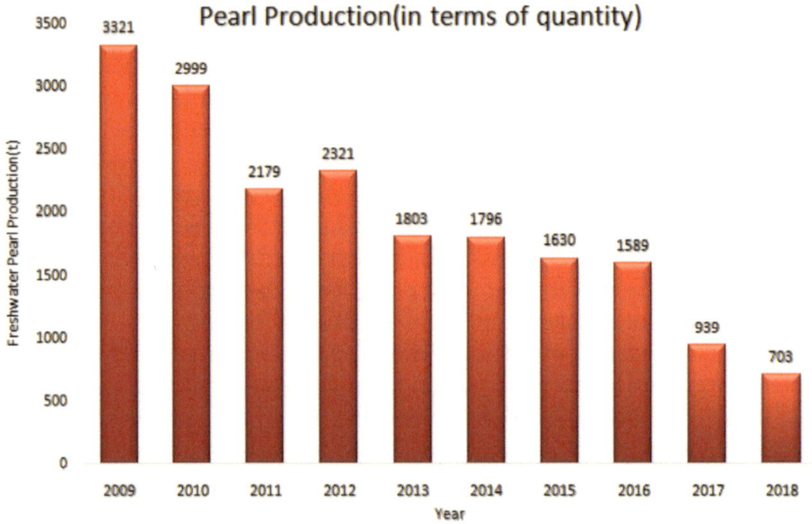

Figure 1.
Global freshwater pearl production in terms of quantity (2009–2018; FAO 2020). X-axis: The annual year of pearl production; Y axis: Freshwater pearl production in tonnes.

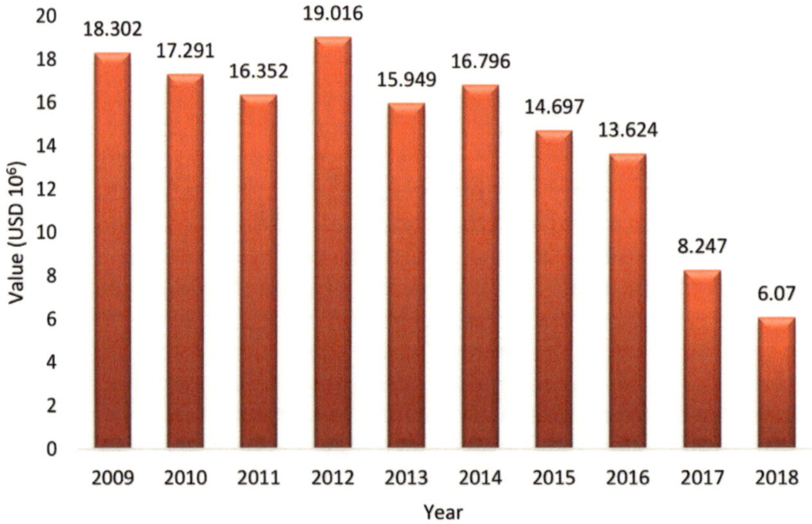

Figure 2.
Total output value of freshwater pearl production (2009–2018; FAO 2020). X-axis: The annual year of pearl production; Y-axis: Value of pearl production.

agencies, export agencies, and others to increase freshwater production output, which will aid in the long-term development of the pearl business. In the marine sector, the pearls from the oyster species *Pinctada maxima* followed by *Pinctada margaritifera* dominate the pearl industry [2].

ICAR-CIFA is instrumental in recognising the importance of freshwater pearl culture in India and creating standardised culture technologies for pearl production in freshwater systems. Collection of mussels, pre-operative care, preparation and implantation of nucleus, post-operative care, culture, and harvest are the six essential phases in the creation of cultured pearls. Of these processes, the precise implantation of nucleus is the most crucial. In comparison to other aquaculture species, pearl cultivation necessitates patience and skill.

2. What is a pearl?

A pearl is a gemstone that is generated by living organisms and is the only one of its kind. Bivalves create pearls as a result of an immunological reaction caused by a foreign particle introduced into their body. The gem is made up of 82–86% of aragonite crystals of calcium carbonate and organic proteins, which form the pearl's matrix, where calcium crystallises [6, 7]. One of the most prevalent proteins identified in pearls is conchiolin. It resembles mother of pearl, the inner nacreous lustrous layer of the shell.

3. Historical development of pearl farming

The English word pearl is derived from the Latin word *pirula* which means pear. The name comes from the fact that pearls are frequently pear-shaped. Since the dawn of time, people have been fascinated by pearls. Pearls were given the term "mukta" in classical Sanskrit, which means "purity" or "escape," alluding to the spirit of the mollusc's desire to escape and solidify as a pearl. Back in the days, procurement of the pearls was done by collecting them from the wild. The creation of these structures in the oyster has long been a source of speculation among ancient philosophers and naturalists. The Rig Veda, the oldest of the Indian Vedas, mentions pearls. Pearl presents are mentioned in Chinese literature dating back to 2200 B.C. Pearls are mentioned in the Indian epics Ramayana and Mahabharata. In Hindu literature, the fabled origin of pearls is linked to Krishna, the eighth avatar of Vishnu, the most significant Hindu divinity. On Pandaia's wedding day, Krishna retrieved pearls from the depths of the sea to adorn her and Indian brides still wear pearls on their wedding days today [8]. According to another narrative, the pearl was a prize for Krishna's victory over the monster Pankagna, and he used it to ornament his bride. In ancient China, it was believed that pearls originated in the brain of the fabled dragon. They were regularly delivered as tributes to emperors by foreign princes and found sparkling in the centre of representations of Gods [8]. Throughout Hebrew literature, the Bible, and the Koran, pearls are mentioned in awe-inspiring terms. King Vijaya, who invaded Sri Lanka in the 6th century B.C., is reported to have lavished pearl offerings to his father-in-law, the Pandyan King of Madurai. Megasthenes, the Greek Ambassador at Chandragupta Maurya's court in the 3rd century B.C., and later travellers like Marco Polo (1260–1300 A.D.) and others have left illustrious records of India's and Sri Lanka's pearl fishery. The Persian Gulf, the Red Sea, and the Gulf of Mannar's pearl bank have all been recognised as producing Oriental pearls. For at least three thousand years, oyster pearl fishing has been documented in the Gulf of Mannar [9]. Pearls reached its zenith in the Roman Empire [10].

In the 12th century A.D., the Chinese found a feasible method for producing pearly Buddha images from the cockscomb pearl mussel, *Cristaria plicata*. There were attempts to make pearls by hand all over the world outside of China. In 1761, Carl von Linnaeus, a Swedish biologist, claimed to have developed a method for producing pearls artificially in freshwater mussels [10]. The Japanese solved the mystery by creating blister pearls from the pearl oyster in 1893 and free spherical pearls from the pearl oyster in 1907. Tokichi Nishikawa and TatsuheiMise brought this perliculture technology to Japan after it was created by British biologist William Saville-Kent in Australia. In 1916, Nishikawa acquired the patent and married Mikimoto's daughter. Later, in Japan, Kokichi Mikimoto, who is regarded as the pioneer of modern pearl cultivation techniques in the nineteenth century, began commercial pearl manufacturing [2], which paved a path towards the present day culture technologies. To distinguish the Japanese pearls from the "natural pearl" produced by wild oysters/mussels, the term "cultured pearl" was coined. Marine pearl farming began in India in the early 1970s, while freshwater pearl culture began in 1989. The marine pearl culture technique was introduced by the ICAR-Central Marine Fisheries Research Institute, whereas the freshwater pearl culture technology was developed by the ICAR-Central Institute of Freshwater Aquaculture. Farmers have expressed a strong desire to participate in the program offered by the ICAR research institute. From 2012 to 2020, more than 2000 entrepreneurs were trained on various elements of freshwater pearl farming through a total of 21 training programmes. A large number of stakeholders have adopted this technology at their backyard as a result of these. Furthermore, a user-friendly manual in various Indian languages has been created to meet the needs of various kinds of aspirants and to reach out to more farmers. Moreover, with the collaboration of ICAR-CIFA scientists, DD Kissan Delhi Doordharsan National Channel, New Delhi produced a documentary film on 'Moti Ke Kheti', which has been highly welcomed by many sections of the Indian population. Later on, DD Kissan released a documentary film on YouTube in order to make freshwater pearl technology accessible to a larger number of people throughout the world.

4. Distribution of freshwater pearl mussels

As per the studies conducted world-wide, the first appearance of pearl producing molluscs precedes to 530 million years [11] with 10,000 species of bivalve reported from across the planet [12]. China harbours more than 100 species of freshwater mussel out of which 10 species belonging to the Unionidae and Margaritiferidae family are being utilised for commercial production of pearls. Few of them include *Hyriopis cumingii, C. plicata, Lamprotula leai, Lamprotula rochechouarti* and *Margaritiana dahurica*. The freshwater pearl market of China is occupied by the pearls obtained from *H. cumingii* followed by *C. plicata* as they offer the ease of operation along with better quality pearls [2]. The freshwater pearl industry of Japan is dominated by *Hyriopis schlegelii* and *Margaritiana dahurica* [2]. The species *H. schlegelii* has strong ability to secrete nacre. Another freshwater mussel species native to North America, *Potamilus alatus*, has the ability to produce high quality black pearls [2]. Due to over-exploitation of European pearl mussel *Margaritifera margaritifera* in search of pearl the species now comes under endangered category and lots of projects are running to revive the natural population of this important species. The distribution of different freshwater pearl mussel is compiled and presented in **Table 1**.

India is a home to around 3270 molluscan species and 1100 out of them are bivalves [62] The marine bivalves count reaches up to 625 species, 88 of which are

Country/Place	Species	References
Bangladesh	*Lamellidens marginalis*	[13–15]
	Lamellidens corrianus	[14]
	Lamellidens jenkinsianus	
	Lamellidens phenchooganjensis	
	Parreysia corrugata	
	Parreysia favidens	
	Parreysia daccaensis	[16]
China	*Cristaria plicata*	[17]
	Hyriopsis cumingii	[18, 19]
	Lamprotula tortuosa	[20]
	Lamprotula leai	[21]
	Lamprotula rochechouarti	[21]
	Lanceolaria glayana	[22]
	Hyriopsis schlegelii	[23]
Czech Republic	*Margaritifera margaritifera*	[24]
Europe	*Margaritifera auricularia*	[25]
	Margaritifera margaritifera	[26]
Finland	*Margaritifera margaritifera*	[27]
France	*Margaritifera margaritifera*	[28]
Germany	*Margaritifera margaritifera*	[29]
India	*Lamellidens marginalis*	[30]
	Lamellidens corrianus	
	Parreysia corrugata	
Indonesia	*Anodonta woodiana*	[31]
Ireland	*Margaritifera margaritifera*	[32]
	Margaritifera durrovensis	
Japan	*Hyriopsis schlegelii*	[33]
	Margaritifera laevis	[34]
	Margaritifera togakushiensis	
	Cristaria plicata	[35]
	Margaritiana dahurica	[2]
Malaysia	*Hyriopsis bialata*	[36]
Mexico	*Psoronaias crocodilurum*	[37]
	Potamilus alata	
Morocco	*Margaritifera marocana*	[38]
Nepal	*Lamellidens marginalis*	[39]
North America	*Quadrula sp.*	[40]
Norway	*Margaritifera margaritifera*	[41]
Philippines	*Cristaria plicata*	[42]
Poland	*Margaritifera margaritifera*	[43]
Portugal	*Margaritifera margaritifera*	[44]

Country/Place	Species	References
Russia	*Margaritifera margaritifera*	[45]
	Margaritifera dahurica	[46]
	Margaritifera middendorfii	
	Margaritifera laevis	
	Cristaria plicata	[47]
Scotland	*Margaritifera margaritifera*	[48]
South Korea	*Cristaria plicata*	[49]
Spain	*Margaritifera auricularia*	[50]
Sweden	*Margaritifera margaritifera*	[51]
Taiwan	*Anodonta woodiana*	[31]
Thailand	*Hyriopsis (Limnoscapha) myersiana*	[52, 53]
	Hyriopsis desowitzi	[52]
	Chamberlainia hainesiana	[52]
	Hyriopsis bialatus	[54]
	Cristaria plicata	[55]
Turkey	*Unio terminalis*	[56]
	Potamida litoralis	
	Leguminaia wheatleyi	
	Anodonta pseudodopsis	
USA	*Quadrula ebena*	[57, 58]
	Quadrula undulate	
	Unio sp.	
	Pleurobema oesopus	
	Tritogonia verrucosa	
	Margaritifera margaritifera	
Vietnam	*Sinohyriopsis cumingii*	[59, 60]
	Cristaria bialata	[61]
	Sinanodonta elliptica	
	Sinanodonta woodiana	
	Lamprotula leai	

Table 1.
Distribution of freshwater pearl mussels in different countries.

endemic [63]. As far as the freshwater mussel species are concerned, around 52 species have been reported from Indian waters including stagnant to slow flowing water bodies [64]. Nevertheless, large scale production of pearls is being carried with three species categorised under the Unionidae family i.e. *L. marginalis*, *L. corrianus* and *Parreysia corrugata* [30, 65].

5. Biomineralisation of pearls

Biomineralisation is a common occurrence in a variety of species [66]. A pearl is a well-known organo-mineral composite product of biomineralisation that is composed of more than 95% calcium carbonate ($CaCO_3$) and less than 5% organic molecules [6, 67, 68]. The process by virtue of which the pearl develops in mussel/oyster is otherwise termed as biomineralisation and the outer epithelium of the mantle plays an important part in pearl biomineralisation. It is formed when an external stimulus like a foreign body or a parasite is fortuitously trapped in the bivalve followed by biomineralisation around the non-native particle via the deposition of pearl nacre in micro-layers. Nevertheless, this phenomenon is restricted to those bivalve species whose interior surface of the shell carries the nacreous layer. The peculiarity of pearl formation being restricted to certain species of bivalves can also be supported by the fact that a prominent resemblance in the $CaCO_3$ and matrix protein component lies between the pearl and nacreous layer of the organism's shell.

Biomineralisation is attributed to a complex physiological process wherein various matrix proteins and calcium metabolism regulatory proteins secreted from the mantle epithelium act together to form the lustrous product that is preceded by the formation of the pearl sac around the foreign particle by the proliferation of the outer epithelium of the mantle [6, 7, 69–71]. The matrix proteins contribute to the pearl formation process in forming a biomineral framework and simultaneously regulating the nucleation as well as the growth of calcium carbonate crystals [6, 7]. Microstructure analysis of pearls reveals that the crystallising layers of the pearl include three polymorphs of $CaCO_3$ that include aragonite, calcite and vaterite, which also govern the quality of pearls [6]. The mantle shell of molluscan species consists of three layers, the outer periostracum, middle prismatic and the inner nacreous layer [7, 72, 73]. The periostracum, a layer of strongly cross-linked proteins that covers the shell's external surface, is produced by the mantle [74]. Secretions from the edge region of the mantle (mantle edge) form the prismatic layer while the inner part of the mantle (mantle core) makes up the nacreous layer or "mother-of-pearl", which is the most widely studied structural motif. Studies on the crystal structures of molluscan shells reveal that the prismatic and nacreous layer consists of calcite and aragonite, respectively [67, 73] and this polymorph formation is regimented by the differences in the mantle epithelium's secreted protein [7, 67]. The third polymorph, vaterite is an unstable and rarely occurring form of $CaCO_3$ that has been exclusively reported in freshwater pearl mussel species [6, 75, 76].

6. Classification of pearls

Pearls can be broadly classified into 3 different types (**Figure 3**). They are as follows-

6.1 Natural pearls

These pearls form naturally in the environment when mussels ingest a foreign particle without the need for human involvement. Natural pearls have a small core or nucleus with thicker crystalline pearl nacre. It has an uneven shape and is quite small. Due to the margins of the overlaying aragonite crystals, the surface of natural pearl has a rough texture.

6.2 Cultured pearls

It's similar to the natural pearls formed in mussels, except instead of an organism accidentally absorbing a foreign particle, a nucleus and a mantle graft are surgically implanted into the mussel. It is possible to create pearls of desired size, shape, colour, and lustre using the culture method of obtaining natural pearls. They come in a variety of shapes and sizes, including round, half-round, and designer pearls, (**Figure 4**) depending on the nucleus utilised.

6.3 Artificial or imitation pearls

Artificial pearls are made by covering a hard, spherical core or base with pearl-like materials to substitute natural or cultured pearls. From low-cost gleaming paints to synthetic pearl essences derived from fish scales, the coating can indicate a difference. Artificial pearls, unlike natural or cultured pearls, have a smooth surface texture and leave a scratch on the surface when rubbed against a sharp object figure depicts the 3 types of pearls and not artifical pearl, only.

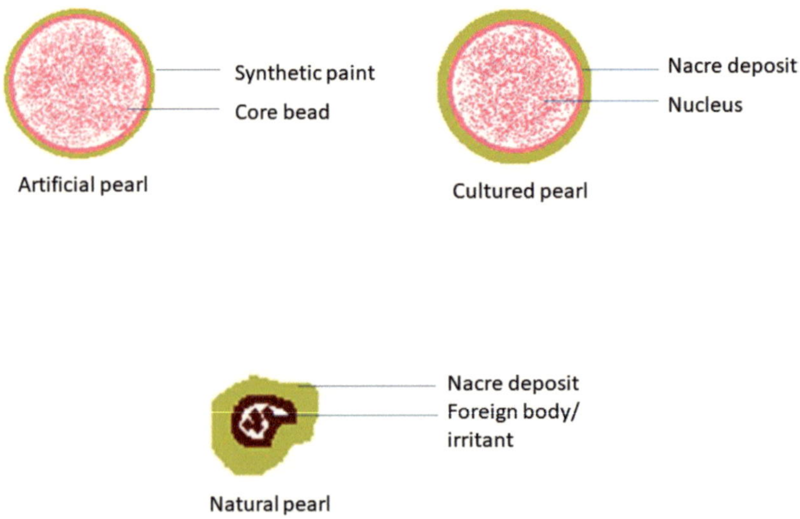

Figure 3.
Graphics depicting distinct types of freshwater pearls.

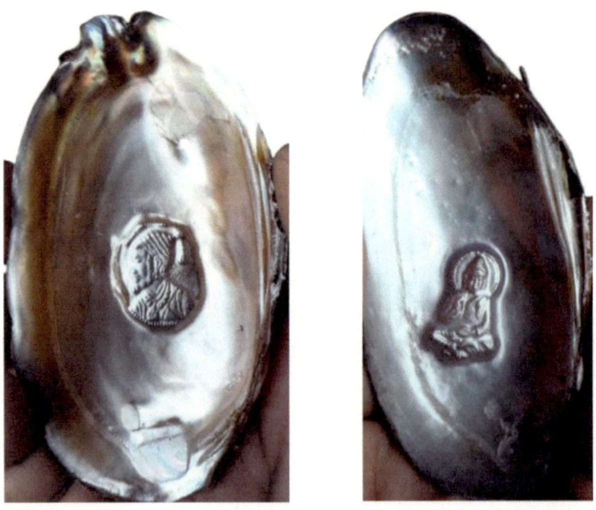

Figure 4.
Production of designer pearl in L. marginalis.

7. Freshwater pearl farming technology

Freshwater pearl farming procedures vary depending on the surgery performed on the pearl mussel's internal anatomy and the type of ideal pearl result. Cultured pearl production involves six basic steps: mussel collection, pre-operative conditioning, nucleus preparation and surgical insertion, post-operative care, pond culture of implanted mussels, and harvest. To avoid graft and mantle rejection and minimise the post-operative mortality rate, implantation can be done at any time of year, with the exception of May–June (summer season). The different steps involved in pearl farming are given below:

7.1 Collection of mussels

The study was conducted at ICAR-CIFA, Kausalyaganga farm, Bhubaneswar, Odisha, India (Lat. 20° 11′ 06″-20° 11′ 45″N; Long.85° 50′52″-85° 51′35″E). Mussels are hand-picked and harvested by competent employees from culture pond, ideally early in the morning. The mussels are harvested and stored in water for short distance travels to avoid stress during transportation. While collecting mussels, the mussel size requirement is the most significant consideration. It has been proven that pearl mussels with a shell length of 8–12 cm and a weight of 35 g or more can produce pearls.

7.2 Pre-operative conditioning

Pre-operative conditioning allows appropriate relaxation of the adductor muscles and keeps the animal's metabolic rate low to get the best possible result throughout the surgical implantation process. Prior to surgery, pearl mussels are subjected to pre-operative conditioning for 24 to 48 hours after being collected from cultured ponds. The mussel was stocked in 200 litre FRP tanks at 60 mussels per tank during acclimation. The experimental tank was filled with filtered pond water. Green water should be offered to them as a source of nutrients. During the study, the following physico-chemical water quality parameters were measured: dissolved oxygen 5.6 to 6.5 mg l^{-1}; pH 7.5 to 7.6; nitrites 0.020 to 0.030 mg l^{-1}; ammonia 0.05 to 0.08 mg l^{-1} and temperature 27 to 29°C. Further, 5% mortality of the mussel was also observed during this period.

7.3 Preparation and surgical implantation of nuclei

It is the most crucial stage in the pearl-growing process. The cultivated pearl design that will be used as the end result will define the type of nucleus to be developed. Some of the nuclei that can be made are designer, half-round, and round nuclei. The nucleus can be prepared using one of two methods. The first step involves mixing acrylic powder with its solvent in a 1:1 ratio to make slurry that is utilised to prepare all of the nucleus forms. The second method, on the other hand, comprises a time-consuming technique that involves bleaching, powdering, and sieving dead mussel shells to get a finely crushed shell powder. The shell powder is then mixed with araldite glue to make dough, which is then cast on moulds to make designer and half-round nuclei, or rolled on the palm to make circular nuclei. Prior to either the acrylic powder-solvent slurry or the shell powder-araldite dough to the moulds, make sure they are lubricated to make it easier to remove the nuclei from the moulds.

The ultimate pearl to be generated determines the method of implantation to be used. There are three different kinds of implantation procedures. The list is as follows:

7.3.1 Mantle cavity implantation

This is the easiest way of the three, requiring the least amount of skill and expertise. The nucleus is implanted into the cavity between the outer mantle layer and the mussel shell's inner surface in this procedure. The outer mantle layer serves as the source of nacre secretion; hence mantle graft is not employed in this procedure.

7.3.2 Mantle tissue implantation

The nucleus, together with the mantle graft, is implanted into pockets formed in the recipient mussel's mantle tissue on the posterior side, in both the left and right lobes (**Figure 5**).

7.3.3 Gonadal implantation

A small incision is made in the recipient mussel's gonad and the nucleus together with a single mantle graft or sandwiched between two pieces of graft are then put into the incision. Along with the round nucleus, a live graft of 2 to 3 mm is introduced from the pallial mantle ribbon. Sometimes, infections can occur in the mussel after the graft and nucleus have been implanted; these infections can be treated with broad-spectrum antibiotics, which also assist to restore the immunological status of the operated mussel.

The success rate of pearl generation in the mantle cavity and mantle tissue implantation methods is 60–70 percent, but it is 25–30 percent in gonadal implantations.

7.4 Post-operative care

The implanted mussels' recuperation necessitates post-operative care, which is a key phase in the production of freshwater pearls. In this step, 7–10 days post the implantation the mussels are monitored closely. The shell valves are given enough care to guarantee that they can open and close freely for respiration. The use of a broad-spectrum antibiotic in the water in post-operative care units at a rate of 1–2 ppm is favourable to the survival and wound healing of the implanted mussels. Mussels in post-operative care are given green algae grown in the lab. During post-operative care, there is a risk of rejection of the implanted nucleus and graft, which can be reduced by lowering stress in the post-operative care tanks by maintaining water level, feeding, and aeration. It is vital to keep in mind that proper post-operative care reduces the likelihood of abnormal pearls by preventing nucleus extrusion soon after implantation.

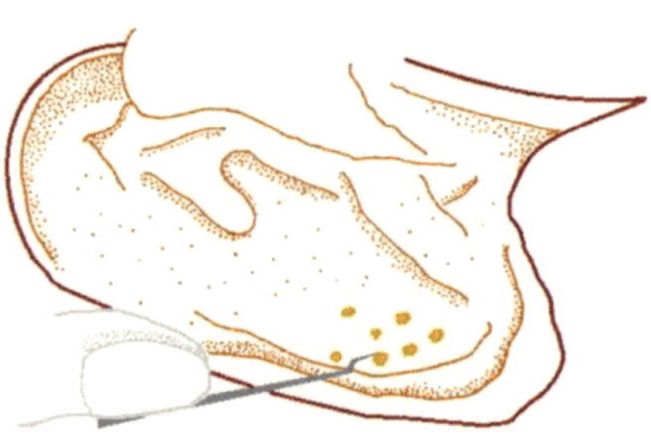

Figure 5.
Implantation of live graft pieces into the mantle of mussel.

7.5 Culture in ponds

To prevent post-operative mussel mortality and nucleus bead rejection, freshwater mussel implantation is done all year in India, except during the extreme hot months (May to June). Pearl culture operations can be carried out in traditional carp culture ponds (2.5 m deep) with a clay-soil basis and somewhat alkaline waters. Ponds free from aquatic macrophytes and algal blooms, such as *Microcystis* and *Euglena*, are perfect for pearl culture. Bamboo poles are used as rafts in the ponds to hang the pearl mussels that have been implanted at a stocking density of 50,000/ha. The mussels are hanged in 1.5 mesh sized nylon bags of 30 cm x 13 cm @ 2 mussels per bag.

Pond management is crucial during the culture stage, especially in terms of natural food production and water quality control via liming or fertilisation, as it affects the quality and quantity of pearl production. The addition of green water (*Chlorella sp., Chlorococcum sp., and Scenedesmus sp.*) to the pearl culture ponds at regular intervals as direct mussel feed has been revealed to be the most effective technique for maintaining the pearl yielding mussel standing crop. The "open culture method" can be used to cultivate green algae in ferro-cement tanks (200 litres) positioned along pond dykes. Cow dung (10,000 kg/ha/yr), urea (100 kg/ha/yr), and single super phosphate (100 kg/ha/yr) are used to fertilise the water in the tanks in equal monthly instalments. When the fertilisers degrade in 10 to 15 days and green water appears, the improved water is placed into the pearl culture ponds. According to reports, freshwater mussels feed on algae belonging to the Chlorophyceae (green algae), Bacillariophyceae (diatoms), and Cyanophyceae (blue green algae) families. Diatoms, green algae (*Chlorella, Chlorococcum, Scenedesmus*, etc.) and blue-green algae (*Spirulina*) are the most usually favoured algal species by the freshwater mussel *L. marginalis*. Mussels may ingest a wide variety of particulate organic materials since they are mucoid filter feeders. A routine health check-up of the cultured mussels should be done at biweekly intervals because there is still a high risk of mortality of operated mussels due to internal incision, limited food availability, and parasitic infection. As a result, mussels in net bags should be removed, inspected, and cleaned before being returned. Because the mussels remain sedentary and static inside the enclosures, algal development can occur as a result of heavy nutrient loading, which should be avoided at all costs. The pond's physico-chemical parameters and water level are monitored during the culture period. Temperatures between 25 and 30 degrees Celsius are optimal.

7.6 Harvest of pearls

Depending on the implantation method utilised, the size and quantity of nuclei implanted, the health of the mussels, and the pond environment, pond culture of operated mussels can take anywhere from a year to 18 months. Harvesting occurs at the end of the culture period, where the mussels are checked and processed individually in order to obtain the ultimate product, pearls. A biological system produces pearls through a natural process. Because pearls are created through a natural process, they have a wide range of look and quality. After harvest, the pearls are subjected to value addition through surface cleaning, bleaching and dyeing, or both cleaning and bleaching, in order to maintain uniformity in colouring and quality, which may improve their marketability.

8. Pearl quality, factors affecting and quality enhancement

The value of a pearl, like any other valuable gem, is decided by its quality, which is decided by several qualities of the pearl such as shape, size, colour, lustre, and surface complexion [69, 77]. Given that a pearl is the result of a complex biological phenomenon, diversity in pearls is unavoidable and is influenced by both the genetic makeup of the individual and the impact of many environmental factors. It can be deduced that genetics, environment, and genotype-by-environment interactions influence the overall quality of pearls [78, 79]. Pearls are classified into three varieties based on their appearance to the naked eye: nacreous pearls, prismatic pearls, and organic pearls, with nacreous pearls being the most valuable [80]. Although controlling and regulating the quality of cultured pearls at the genetic level is a time-consuming task that necessitates extensive research, changes in culture methods and environmental conditions can significantly improve pearl quality. Pearl quality is also affected by other factors like the host and donor oysters [70], or by exogenous physico-chemical and biological stressors [81]. Various factors that can be regulated to enhance the pearl quality are listed below.

8.1 Host

Individual mussels selected for surgical nucleus implantation must be healthy and of a suitable size.

8.2 Donor mussel and graft tissue preparation

The donor mussel used to get the graft tissue should have a well-developed and healthy mantle and be of the desired size. The mantle graft should be carefully selected, cleaned, cut, and trimmed, and the graft tissues should be kept in good quality water with the proper level of chemicals while the surgical procedures are being performed.

8.3 Implantation

The most important phase in the production of cultured pearls is nucleus implantation. A successful implantation technique is defined, in addition to the technician's skills, by the identification of an appropriate location for nucleus and graft insertion. More skill and patience are necessary to appropriately position and orient the graft tissue in contact with the nucleus, as well as for several nucleus implantations in a single individual.

8.4 Mussel convalescence

For the mussels to recover from the effects of narcotisation, they need regular water changes or a gentle flow through. Furthermore, the mussels should be given ample time to heal after the incision for nucleus implantation before being stocked in culture ponds.

8.5 Tool maintenance

Before usage, surgical instruments should be sharp, rust-free, and thoroughly sterilised.

8.6 Temperature

Temperature is known to influence the metabolic rate of every organism. Although higher temperatures encourage mussel growth and nacre deposition, the quality of the pearls produced suffers as a result.

8.7 Quantity and quality of natural feed

The type and amount of plankton supplied to the mussels during the culture stage is also a quality determining factor because quality and quantity of nacre secretion of the organism depends on plankton composition.

8.8 Culture period

The mussels must be nurtured for the requisite period of time to develop the proper thickness of nacre surrounding the nucleus, which contributes to the pearl's overall size, colour, and shine. The culture period might last anywhere from 12 to 18 months, depending on the type of nucleus used and implantation employed.

The raw pearl obtained from the implanted freshwater mussels are inapt be to directly used as a jewel. The pearls are value added after harvesting in order to improve quality and maintain consistency. They are subjected to a variety of procedures, including cleaning, bleaching, dyeing, or both. For cleaning or bleaching, hydrogen peroxide, ether solvents, water, and alcohol are used in varied amounts depending on the need. Ultra sonication, for example, is a physical treatment process that removes any clinging contaminants. Other chemicals, such as EDTA, sodium hypochlorite (NaOCl), and calcium hypochlorite (bleaching powder), can help remove adherent particles by chelating, oxidising, or bleaching them. Pearls have been demonstrated to shine when treated with EDTA, NaOCl, chlorine, and hydrogen peroxide [82].

Studies have shown that the colour of cultured pearl can be enhanced using radiation and chemical treatment. 48 h exposure to 5.423% 10^{-2} M rad of gamma radiation, treatment in 1.2% eosin solution for 24 h, 20% iodine solution for 48 h and 0.2% silver nitrate solution for 24 h result in distinct changes in the colour of pearl from white to black, pink, yellow and metallic brown or black, respectively. It was also observed that the coloration caused by gamma radiation and silver nitrate was everlasting [82].

9. Identification of pearl

There are a few characteristics that can be used to establish the originality of a pearl, and they are as follows:

9.1 Irregularities

Minor irregularities in colour signify a pearl to be original in contrast to which artificial pearls are large, symmetrical and perfectly matched in all possible way.

9.2 Examination of drill hole

You can detect if a pearl is natural or not by poking a small hole in it and looking at it under a magnifying glass. The crystallisation layers of a real pearl can be seen using a magnifying lens.

9.3 Tooth test

To identify between genuine and imitation pearls, gently rubbing the pearl between the teeth is a frequent and fairly reliable test. A natural pearl's surface has a gritty and sandy feel due to the unique architecture of the nacreous surface, but an artificial pearl has a smooth texture and hence a smooth feel when rubbed on teeth.

9.4 Lustre test

An artificial pearl's lustre is limited to its surface, and scratching it against a rough surface removes the shiny coating. A genuine pearl, on the other hand, is formed by the subsequent deposition of nacre in layers, thus the lustrous nacre layers run deep from the surface up to the nucleus, and scratching the pearl surface is difficult due to the compactness of the calcium carbonate crystallisation.

9.5 Spectroscopy

Micro-infrared spectroscopy, Raman spectrometry, scanning electron microscopy and X-ray diffraction are all commonly employed to distinguish between freshwater and saltwater cultured pearls [83].

10. Grading of pearls

Natural cultured pearls that are commercially accessible are graded according to their quality [84], as shown below.

Grade	Remarks
AAA	Excellent lustre, no surface flaws, and good symmetry characterise the highest grade with superb attributes
AA	Good quality, good lustre, homogeneous coloration with a few surface defects
A	Medium quality, good lustre, non-uniformity in colouration with some surface imperfections and poor symmetry
B	Good lustre with irregular surface and coloration, as well as a few surface flaws
NC	No economic value, the shine is low, the nacre layer is weak, and serious flaws on the surface

11. Challenges in freshwater pearl culture

Despite the fact that pearl farming is a profitable business, there are a number of challenges to overcome when cultivating the pearl. One of the most crucial aspects is the mussel's capacity to survive following implantation. Another challenge is determining the right quality of the pearl after it has been obtained. In freshwater mussels, standardisation of breeding technology needs an utmost priority, successful breeding occurs in mussels; however mussel larvae survival is a serious challenge. i.e., glochidia adhesion to the secondary host (fish) is problematic. The mussel larval cycle requires a secondary host, such as a fish, to complete the life cycle. The lack of competence in pearl culture technology is one of the primary challenges. The sector's development is further hampered by a lack of a proper extension network to propagate existing culture technology.

12. Prospects of freshwater pearl culture in India

Farmers in developing countries like India have a limited grasp of modern aqua farming procedures, including pearl farming, which should be followed in their respective fields. Through research, teaching, and training, many people are working hard to transfer this critical technology of pearl farming to the needy. Many farmers, entrepreneurs, and women who are interested in this subject have already received training in recent years. Freshwater pearl farms have been developed in various states across the country, including Odisha, Maharashtra, Gujarat, West Bengal, Bihar, Uttar Pradesh, Chhattisgarh, Kerala, and a few more that are still in the early phases of development with the technical support of ICAR-CIFA. Larger round pearls and designer pearls in various designs such as goddess Laxmi, Holy Cross, Ganesha, and other beautiful shapes are currently in high demand in India. Time has come to educate more people about pearl farming, as this aquaculture technology is expected to generate a lot of employment and money. To produce excellent pearls in a shorter period, new developments in freshwater pearl farming should be introduced into farmer's fields in the future.

13. Conclusion

The ICAR-Central Institute of Freshwater Aquaculture has made significant progress in areas such as the identification of newer biocompatible nuclei, surgical implantation technique, mussel pre and post-operative care, graft and nucleus rejection minimization, pond culture of implanted mussels, and pearl value addition. Efforts are also being made to develop low-cost technologies for farmers through the use of basic, readily available equipment in pearl cultivation. Glochidial larvae culture *in vivo* and *in vitro* has also been the bailiwick of recent research. The technology of freshwater pearl culture is also being disseminated with a focus on farmers, investors, state government officials, researchers, and students across the country to enhance technical skills and popularise freshwater pearl culture technology to new heights. However, the limited availability of skilled personnel for precise implantation and a lack of effective marketing networks are the two most significant barriers to successful pearl farming adoption. Commercialization of this skilled technology necessitates a well-thought-out strategy at the regional and farm levels, as well as entrepreneurial development and participation. To summarise, freshwater pearl farming has a large economic, social, and environmental impact. In the future, this pearl-growing method is expected to become one among the most well-known components of India's freshwater aquaculture.

Acknowledgements

The authors wish to thank the Director, ICAR-Central Institute of Freshwater Aquaculture for providing necessary facilities for carrying out the work.

Conflict of interest

The authors declare no conflict of interest.

Author details

Shailesh Saurabh[1*], Sweta Pradhan[1] and Sonal Suman[1,2]

1 ICAR-Central Institute of Freshwater Aquaculture, Kausalyaganga, Bhubaneswar, Odisha, India

2 ICAR-Central Institute of Fisheries Education, Mumbai, India

*Address all correspondence to: ssaurabh02@rediffmail.com

References

[1] Alagarswami K. Cultured pearls-production and quality. CMFRI Bulletin-Pearl culture. 1987; 39:107-111.

[2] Zhu C, Southgate PC, Li T. Production of pearls. InGoods and services of marine bivalves 2019 (pp. 73-93). Springer, Cham.

[3] Gervis MH, Sims NA. The biology and culture of pearl oysters (Bivalviapteriidae). WorldFish; 1992.

[4] Southgate PC, Strack E, Hart A, Wada KT, Monteforte M, Cariño M, Langy S, Lo C, Acosta-Salmon H, Wang A. Exploitation and culture of major commercial species. The pearl oyster. 2008;303:355.

[5] FAO. Fisheries and Aquaculture Statistics. Food and Agriculture Organization of the United Nations, Rome; 2020. 248p.

[6] Jin C, Li J. The Molecular Mechanism of Pearl Biomineralization. Annals of Aquaculture and Research. 2017.

[7] Kinoshita S, Wang N, Inoue H, Maeyama K, Okamoto K, Nagai K, Kondo H, Hirono I, Asakawa S, Watabe S. Deep sequencing of ESTs from nacreous and prismatic layer producing tissues and a screen for novel shell formation-related genes in the pearl oyster. Plos one. 2011 Jun 22;6(6):e21238.

[8] Kunz GF, Stevenson CH. The book of the pearl: its history, art, science and industry. Courier Corporation; 2013 Feb 19.

[9] Barroso MD. Pearl, An Ancient Antidote of Eastern Origin. In Toxicology in Antiquity 2019 Jan 1 (pp. 401-410). Academic Press.

[10] Alagarswami K. Production of cultured pearls. Published by Indian Council of Agricultural Research, New Delhi; 1991. 112p.

[11] Pearl Knowledge. Available from: https://rawpearls.com.au/pearl_knowledge [Accessed: 2021-06-17]

[12] The Bivalvia. Availablefrom: https://ucmp.berkeley.edu/taxa/inverts/mollusca/bivalvia.php [Accessed: 2021-06-17]

[13] Miah MI, Rahman AS, Rahmatullah SM, Saha JK, Islam MA. Culture of pearl in freshwater mussels (Lamellidens marginalis Lamarck). Bangladesh Journal of Fisheries Research. 2000;4(1):57-61.

[14] Hossain MA, Sultana N, Azimuddin K, Hussain MG, Mazid MA. Selection of freshwater pearl mussel species for mantle transplantation in Bangladesh. Bangladesh Journal of Fisheries Research. 2004;8(2):113-116.

[15] Niogee SR, Tonni KF, Barman AC, Tanu MB, Sku S, Uddin MJ. Ovarian Cycle of Freshwater Pearl Mussel, *Lamellidens marginalis* (Lamarck, 1819) Collected from a Culture Pond in Bangladesh. Asian Fisheries Science 2019:32:117-123.

[16] Pagcatipunan R. Technical assistance on oyster and pearl culture in Bangladesh. 1984. FAO. http://www.fao.org/3/r2032e/R2032E00.htm#TOC

[17] Dan H, Ruobo G. Freshwater pearl culture and production in China. Aquaculture Asia. 2002;7(1):6-8.

[18] Bai ZY, Li JL, Wang GL. Relationship between pearl production, growth traits and the inserted position of mantle piece in triangle mussel (Hyriopsis cumingii). J Fish Sci China. 2008;15(3):493-499.

[19] Lin JY, Ma KY, Bai ZY, Li JL. Molecular cloning and characterization

of perlucin from the freshwater pearl mussel, Hyriopsis cumingii. Gene. 2013 Sep 10;526(2):210-216.

[20] Wang G, Cao X, Li J. Complete F-type mitochondrial genome of Chinese freshwater mussel Lamprotula tortuosa. Mitochondrial DNA. 2013 Oct 1;24(5):513-515.

[21] Bai Z, Wang G, Liu X, Li J. The status and development trend of freshwater pearl seed industry in China. Journal of Shanghai Ocean University. 2014;23(6):874-881.

[22] Wang G, Chen M, Li J. Complete F-type mitochondrial genome of freshwater mussel Lanceolaria glayana. Mitochondrial DNA Part A. 2016 Mar 3;27(2):846-847

[23] Wu D, Wang C, Zhang W, Peng K, Sheng J, Wang J, Jain A, Hong Y. Molecular characterization of an inhibitor of apoptosis protein (IAPs) in freshwater pearl mussel, Hyriopsis schlegelii. Bioengineered. 2019 Jan 1;10(1):365-373.

[24] Simon OP, Vaníčková I, Bílý M, Douda K, Patzenhauerová H, Hruška J, Peltánová A. The status of freshwater pearl mussel in the Czech Republic: several successfully rejuvenated populations but the absence of natural reproduction. Limnologica. 2015 Jan 1;50:11-20.

[25] Prié V, Soler J, Araujo R, Cucherat X, Philippe L, Patry N, Adam B, Legrand N, Jugé P, Richard N, Wantzen KM. Challenging exploration of troubled waters: a decade of surveys of the giant freshwater pearl mussel Margaritifera auricularia in Europe. Hydrobiologia. 2018 Mar;810(1):157-175.

[26] Sousa R, Ferreira A, Carvalho F, Lopes-Lima M, Varandas S, Teixeira A, Gallardo B. Small hydropower plants as a threat to the endangered pearl mussel Margaritiferamargaritifera. Science of the Total Environment. 2020 Jun 1;719:137361.

[27] Oulasvirta P, Leinikki J, Syväranta J. Freshwater pearl mussel in Finland—current status and future prospects. Biology Bulletin. 2017 Jan;44(1):81-91.

[28] Legalle M, Mastrorillo S, Céréghino R. Spatial distribution patterns and causes of decline of three freshwater species with different biological traits (white-clawed crayfish, bullhead, freshwater pearl mussel): a review. InAnnales de Limnologie-International Journal of Limnology 2008 (Vol. 44, No. 2, pp. 95-104). EDP Sciences.

[29] Jungbluth JH. Conservation projects for the freshwater pearl mussel Margaritifera margaritifera in the Federal Republic of Germany. InThe Conservation Biology of Molluscs: Proceedings of a Symposium Held at the 9th International Malacological Congress, Edinburgh, Scotland, 1986 1995 (No. 9, p. 34). IUCN.

[30] Janakiram K. Freshwater pearl culture technology development in India. Journal of Applied Aquaculture. 2003 Jun 1;13(3-4):341-349.

[31] Rahayu SY, Solihin DD, Manalu W, Affandi R. Nucleus pearl coating process of freshwater mussel Anodonta woodiana (Unionidae). HAYATI Journal of Biosciences. 2013 Mar 1;20(1):24-30.

[32] Geist J, Moorkens E, Killeen I, Feind S, Stoeckle BC, Connor ÁO, Kuehn R. Genetic structure of Irish freshwater pearl mussels (Margaritifera margaritifera and Margaritifera durrovensis): Validity of subspecies, roles of host fish, and conservation implications. Aquatic Conservation: Marine and Freshwater Ecosystems. 2018 Aug;28(4):923-933.

[33] Shirai A, Kondo T, Kajita T. Molecular markers reveal genetic contamination of endangered freshwater pearl mussels in pearl culture farms in Japan. Venus: the Japanese journal of malacology. 2010;68(3):151.

[34] Takeuchi M, Okada A, Kakino W. Phylogenetic relationships of two freshwater pearl mussels, Margaritifera laevis (Haas, 1910) and Margaritifera togakushiensis Kondo & Kobayashi, 2005 (Bivalvia: Margaritiferidae), in the Japanese archipelago. Molluscan Research. 2015 Oct 2;35(4):218-226.

[35] Sano I, Shirai A, Kondo T, Miyazaki JI. Phylogenetic relationships of Japanese Unionoida (Mollusca: Bivalvia) based on mitochondrial 16S rDNA sequences. Journal of water resource and protection. 2017 Apr 17;9(5):493-509.

[36] Razak NF, Supramaniam CV, Zieritz A. A dichotomous PCR–RFLP identification key for the freshwater mussels (Bivalvia: Unionida) of Peninsular Malaysia. Conservation Genetics Resources. 2019 Dec;11(4):457-464.

[37] Saucedo PE, Acosta-Salmón H, McLaurin-Moreno D, Castillo-Domínguez A, Melgar-Valdés CE, Mazón-Suástegui JM. Freshwater pearl culture in Mexico: historic context, present status and future perspectives. Reviews in Aquaculture. 2021 Jun 1;13:1379-1396.

[38] Sousa R, Varandas S, Teixeira A, Ghamizi M, Froufe E, Lopes-Lima M. Pearl mussels (Margaritifera marocana) in Morocco: Conservation status of the rarest bivalve in African fresh waters. Science of the Total Environment. 2016 Mar 15;547:405-412.

[39] Husen, A., Gurung, T.B., Nepal, A. P. (2018). Freshwater pearl culture: an initiative in Nepal. In: Proceedings of the 2nd NEFIS International Convention on "Sustainable Fisheries & Aquaculture Diversification" (Ed. N.P. Pandit and N. Pradhan), Nepal, pp. 137-144.

[40] Graf DL, FOIGHIL DÓ. The evolution of brooding characters among the freshwater pearly mussels (Bivalvia: Unionoidea) of North America. Journal of Molluscan Studies. 2000 May 1;66(2):157-170.

[41] Marwaha J, Jakobsen PJ, Karlsson S, Larsen BM, Wacker S. Higher mortality of the less suitable brown trout host compared to the principal Atlantic salmon host when infested with freshwater pearl mussel (Margaritifera margaritifera) glochidia. Parasitology Research. 2021 Apr 12:1-3.

[42] Battad EM. Polyculture of the freshwater mussel, Cristaria plicata (leach) and Crucian carp, Carrassius carassius in paddies with and without rice [Philippines]. CLSU [Central Luzon State University] Scientific Journal (Philippines). 1984.

[43] Zajac K, Zajac T. The pearl mussel Margaritifera margaritifera (Linnaeus, 1758)(Bivalvia: Margaritiferidae) in Poland-current situation. Folia Malacologica. 2014;22(3).

[44] Sousa R, Amorim Â, Froufe E, Varandas S, Teixeira A, Lopes-Lima M. Conservation status of the freshwater pearl mussel Margaritifera margaritifera in Portugal. Limnologica. 2015 Jan 1;50:4-10.

[45] Popov IY, Ostrovsky AN. Survival and extinction of the southern populations of freshwater pearl mussel Margaritifera margaritifera in Russia (Leningradskaya and Novgorodskaya oblast). Hydrobiologia. 2014 Sep;735(1): 161-177.

[46] Bolotov IN, Bespalaya YV, Vikhrev IV, Aksenova OV, Aspholm PE, Gofarov MY, Klishko OK, Kolosova YS, Kondakov AV, Lyubas AA, Paltser IS.

Taxonomy and distribution of freshwater pearl mussels (Unionoida: Margaritiferidae) of the Russian Far East. PLoS One. 2015 May 26;10(5):e0122408.

[47] Klishko OK, Lopes-Lima M, Froufe E, Bogan AE, Abakumova VY. Systematics and distribution of Cristaria plicata (Bivalvia, Unionidae) from the Russian Far East. ZooKeys. 2016(580):13.

[48] Hastie LC. Are Scottish freshwater pearl mussel populations recruiting normally?. Toxicological and Environmental Chemistry. 2011 Oct 1;93(9):1748-1763.

[49] Patnaik BB, Wang TH, Kang SW, Hwang HJ, Park SY, Park EB, Chung JM, Song DK, Kim C, Kim S, Lee JS. Sequencing, de novo assembly, and annotation of the transcriptome of the endangered freshwater pearl bivalve, Cristaria plicata, provides novel insights into functional genes and marker discovery. PLoS One. 2016 Feb 12;11(2):e0148622.

[50] Araujo R, Ramos MA. Margaritifera auricularia (Unionoidea, Margaritiferidae), the giant freshwater pearl mussel rediscovered in Spain. Graellsia. 1998;54(54):129-130.

[51] Henrikson L, Arvidsson B, Österling M. Aquatic Conservation with Focus on Margaritifera margaritifera: Proceedings of the International Conference in Sundsvall, Sweden, 12-14 August, 2009. Karlstadsuniversitet.

[52] Panha S, Kosavititkul P. Mantle transplantations in freshwater pearl mussels in Thailand. Aquaculture International. 1997 May;5(3):267-276.

[53] Kovitvadhi S, Kovitvadhi U, Sawangwong P, Machado J. A laboratory-scale recirculating aquaculture system for juveniles of freshwater pearl mussel Hyriopsis (Limnoscaphfa) myersiana (Lea, 1856). Aquaculture. 2008 Mar 31;275 (1-4):169-177.

[54] Supannapong P, Pimsalee T, Teerasak A, Engkagul A, Kovitvadhi U, Kovitvadhi S, Rungruangsak-Torrissen K. Digestive enzymes and in-vitro digestibility of different species of phytoplankton for culture of the freshwater pearl mussel, Hyriopsis (Hyriopsis) bialatus. Aquaculture International. 2008 Oct;16(5): 437-453

[55] Chartchumni B, Kumla S, Rangsiwiwat A, Rayan S. Effect of Sizes on Acceptance of Implantation Tissue in Freshwater Mussel Cristaria plicata for Non-Nucleated Pearl Production. Burapha Science Journal. 2020 Sep 1;25(3):1163-1171.

[56] Şerefli̇şan H. Comparison of Pearl Sac Formation in Four Mussel Species (Mollusca: Bivalvia: Unionoida) at the Graft Implantation. Turkish Journal of Agriculture-Food Science and Technology. 2019 Oct 12;7(10): 1699-1704.

[57] Simpson CT. The classification and geographical distribution of the pearly fresh-water mussels. Proceedings of the United States National Museum. 1896.

[58] Federman D. Modern Jeweler's Consumer Guide to Colored Gemstones. Springer Science & Business Media; 2012 Dec 6.

[59] Van Phuc P, Viet PQ, Hoang NM, Tam NT, Ngoc PK. Research on in vitro culture and inducing nacre crystal formation of freshwater pearl mussel mantle epithelial cell Sinohyriopsis cumingii. International Journal of Fisheries and Aquaculture. 2011 Jun 30;3(6):105-113.

[60] Hoang T, Kiet HG, Quang H. Culture and exploration on in vitro explant and instigating nacre gem

development of freshwater pearl mussel mantle epithelial cell Sinohyriopsis cumingii. Advances in Fishery, Aquaculture and Hydrobiology. 2016; 4(1):8-16.

[61] Van Tu Do LQ, Bogan AE. Freshwater mussels (bivalvia: unionida) of Vietnam: diversity, distribution, and conservation status. Freshwater Mollusk Biology and Conservation. 2018;21:1-18.

[62] Mohamed KS, Venkatesan V. Marine molluscan diversity in India–Exploitation, conservation. 2017

[63] Mohamed KS, Sasikumar G. Overview of bivalve fisheries of India. 2016

[64] Subba Rao NV. Handbook of Freshwater Molluscs of India. Zoological Survey of India, Calcutta, 1989. pp. 289.

[65] Saurabh S, Mohanty UL, Mohanty J, Jayasankar P. Pearl culture technology in freshwater environment. Aquaculture and Fisheries Environment. Discovery Publishing House PVT.LTD; New Delhi, India. 2014. pp. 51-78.

[66] Bäuerlein E, Behrens P, Epple M. Handbook of biomineralization. Wiley-VCH; 2007.

[67] Funabara D, Ohmori F, Kinoshita S, Koyama H, Mizutani S, Ota A, Osakabe Y, Nagai K, Maeyama K, Okamoto K, Kanoh S. Novel genes participating in the formation of prismatic and nacreous layers in the pearl oyster as revealed by their tissue distribution and RNA interference knockdown. PLoS One. 2014 Jan 15;9(1):e84706.

[68] Shi Y, Yu C, Gu Z, Zhan X, Wang Y, Wang A. Characterization of the pearl oyster (Pinctada martensii) mantle transcriptome unravels biomineralization genes. Marine biotechnology. 2013 Apr 1;15(2):175-187.

[69] Bai Z, Zheng H, Lin J, Wang G, Li J. Comparative analysis of the transcriptome in tissues secreting purple and white nacre in the pearl mussel Hyriopsis cumingii. PloS one. 2013 Jan 14;8(1):e53617.

[70] McGinty EL, Zenger KR, Jones DB, Jerry DR. Transcriptome analysis of biomineralisation-related genes within the pearl sac: host and donor oyster contribution. Marine genomics. 2012 Mar 1;5:27-33.

[71] Take S, Igarashi Y, Yoshitake K, Asakawa S, Maeyama K, Nagai K, Watabe S, Kinoshita S. Gene expression profiles at different stages for formation of pearl sac and pearl in the pearl oyster Pinctada fucata. BMC genomics. 2019 Dec;20(1):1-21.

[72] Marin F, Luquet G, Marie B, Medakovic D. Molluscan shell proteins: primary structure, origin, and evolution. Current topics in developmental biology. 2007 Jan 1;80:209-276.

[73] Wang X, Liu Z, Wu W. Transcriptome analysis of the freshwater pearl mussel (Cristaria plicata) mantle unravels genes involved in the formation of shell and pearl. Molecular genetics and genomics. 2017 Apr 1;292(2):343-352.

[74] Nudelman F. Nacre biomineralisation: A review on the mechanisms of crystal nucleation. In Seminars in cell & developmental biology 2015 Oct 1 (Vol. 46, pp. 2-10). Academic Press.

[75] Ma HY, Lee IS. Characterization of vaterite in low quality freshwater-cultured pearls. Materials Science and Engineering: C. 2006 May 1;26(4):721-723.

[76] Wehrmeister U, Jacob DE, Soldati AL, Hager T, Hofmeister W. Vaterite in freshwater cultured pearls

from China and Japan. Journal of Gemmology-London. 2007;30(7/8):399.

[77] Le Luyer J, Auffret P, Quillien V, Leclerc N, Reisser C, Vidal-Dupiol J, Ky CL. Whole transcriptome sequencing and biomineralization gene architecture associated with cultured pearl quality traits in the pearl oyster, Pinctada margaritifera. BMC genomics. 2019 Dec;20(1):1-1.

[78] Jerry DR, Kvingedal R, Lind CE, Evans BS, Taylor JJ, Safari AE. Donor-oyster derived heritability estimates and the effect of genotype× environment interaction on the production of pearl quality traits in the silver-lip pearl oyster, Pinctada maxima. Aquaculture. 2012 Mar 29;338:66-71.

[79] Kvingedal R, Evans BS, Lind CE, Taylor JJ, Dupont-Nivet M, Jerry DR. Population and family growth response to different rearing location, heritability estimates and genotype× environment interaction in the silver-lip pearl oyster (Pinctada maxima). Aquaculture. 2010 Jun 15;304(1-4):1-6.

[80] Inoue N, Ishibashi R, Ishikawa T, Atsumi T, Aoki H, Komaru A. Can the quality of pearls from the Japanese pearl oyster (Pinctada fucata) be explained by the gene expression patterns of the major shell matrix proteins in the pearl sac?. Marine Biotechnology. 2011 Feb;13(1):48-55.

[81] Ito M. Improving pearl quality by grafting and husbandry methods. Aqua Tips. 2009;20(1):1-8.

[82] Maharathy C. Hydrochemical materials beneficiation of freshwater culture pearls. PhD Thesis, Berhampur University, Odisha; 2000.

[83] Monarumit N, Noirawee N, Phlayrahan A, Promdee K, Won-In K, Satitkune S. Identification of high-luster and lusterless freshwater-cultured pearls by X-ray absorption spectroscopy. Journal of Applied Spectroscopy. 2015 Sep;82(4):677-680.

[84] Ruiz-Rubio H, Acosta-Salmón H, Olivera A, Southgate PC, Rangel-Dávalos C. The influence of culture method and culture period on quality of half-pearls ('mabé') from the winged pearl oyster Pteria sterna, Gould, 1851. Aquaculture. 2006 Apr 28;254(1-4):269-274.

Chapter 6

Quaternary Marine Mollusk Associations of the Last Interglacials in North Patagonia (Argentina): Paleoecology and Paleoclimates

M.P. Charó

Abstract

Deposits of different Quaternary marine transgressions are largely exposed in the Argentine north Patagonian littoral (39°15′S–41°02′S), south of the Buenos Aires and north of Río Negro provinces. The malacological associations of 84 sites were studied. Among them, 31 belong to Pleistocene deposits of the interglacials ≥ MIS 9, MIS 7, MIS 5e, 29 to Holocene deposits of the interglacial MIS 1, and 24 sites of modern beaches. These sites yielded 7385 fossils among valves and shells, of 78 species (42 bivalves and 36 gastropods), including 11 micromolluskan species. The record of the bivalves *Crassostrea rhizophorae* in the south of the Buenos Aires Province, and *Anomalocardia brasiliana* (both currently inhabiting lower latitudes), and very likely the gastropod *Tegula atra* (inhabiting today the Pacific Ocean) in the north of Río Negro Province, suggests that interglacials MIS 7, MIS 5e and MIS 1 were warmer than today. However, the associations determined for the studied interglacials have not changed in their composition, but in abundance of species, except for the latitudinal shifts of the three mentioned species, and the presence of cold to temperate water taxa since the MIS 1 in the ecotonal area of the north of Río Negro Province. Changes in the associations of northern Patagonia during the Quaternary derived from global changes (sea surface temperature, salinity, etc.), and the existence of habitat heterogeneity in each of the areas, that enabled the co-existence of different bivalve and gastropod species of the local benthic marine malacofauna.

Keywords: quaternary, mollusks, paleoenvironments, paleoecology, north Patagonia

1. Introduction

1.1 Quaternary in the world

The Quaternary is characterized worldwide by important climate oscillations, with extremes represented by glacial and interglacial periods resulting from temperature variations that caused marked changes in sea level (e.g., [1]). In coastal areas, transgressive-regressive events have generated a sequence of erosion forms (coastal terraces and paleocliffs) and beach deposits that, for different reasons, have

IntechOpen

been protected from degradation processes and are therefore an important testimony of climate changes that have occurred in most recent geologic time [2].

1.2 Glacial and interglacial cycles

In 1941, Milankovitch developed a planetary theory that attributes Quaternary glacial and interglacial cycles to modifications of orbital parameters such as eccentricity (100 Ka), obliquity (41Ka), and precession of the equinoxes (19 Ka).

Since the beginning of the Pleistocene, climatic oscillations would have followed periodic cycles of about 40 Ky that seem to conform the variation cycle of the earth axis. The amplitude of the cycles tended to increase 1.5 My ago, and from 600 Ka the glacial cycles have occurred at intervals of between 80 and 120 Ky (e.g., [3, 4]). This duration of the recent cycles is similar to the period of variation of the eccentricity of the earth orbit, of 100 Ky.

1.3 Marine isotopic stages (MIS)

The different glacial and interglacial events occurred during the Quaternary were differentiated through the marine isotope stages (MIS). These MIS represent alternate cold and warm periods established on the basis of δ180 of benthic foraminifers, obtained from cores of the sea bottom [5]. Emialiani [6] divided the last million years in successive isotopic stages on the basis of the δ180/δ160 relationship. Each isotopic stage represents a glacial period (designated with an odd number) or interglacial (designated with an even number), and reveal the advance and retreat of the ice during the last glaciations.

1.3.1 Interglacials in the world

Globally, the MIS 11 encompasses from 424 to 374 ka. It was a long warm period that reached a global mean sea level of 6 to 13 m above the present one between 410 and 400 ka [2]. Some authors (e.g., [7–9]) consider the MIS 11 as an analogous of the Holocene both in climatic conditions and orbital forcing. According to Ashton et al. [10] the climatic conditions in marine isotopic and ice sheet records include at least two large warm episodes with an intermediate cooling phase. The warm conditions of this interglacial were reflected in different marine and terrestrial communities (e.g., [9, 11, 12]).

MIS 9 encompasses from ca. 330 up to 310 ka and sea level was 3 ± 3 m below the present one [13]. In the Northern Hemisphere, at Henderson Island (24°22'S/128°20'W), the highest sea level recorded in MIS 9 is between 334 ± 4 and 324 ± 3 ka in agreement with the maximum sun insulation of 333 ka [14]. In the western Mediterranean (Spain) the sea surface temperature (SST) and the salinity recorded in this interglacial were similar to those of MIS 7 and MIS 5e [15], whereas other authors (e.g., [16]) suggested that according to paleontological evidence, this interglacial was warmer than MIS 7 and the Holocene, and similar to MIS 5e.

MIS 7 encompasses from ca. 245 up to 190 ka [17] with three temperature maximums [18]. Isotopic data of some deep sea cores suggest that the sea level would not have reached the level of the present cero [19, 20], although other authors suggest values around −18 m [17]. SST of this interglacial was higher than the present one [21–23]. In the European coasts, the marine deposits of MIS 7 record the appearance of the "Senegalese" marine fauna from the African coast, confirming this stage as warm in the Northern Hemisphere (e.g., [24–26]). Similar conditions are observed in the Southern Hemisphere: the SST of the southern Argentine Patagonia (42°–43°S) is proposed to be similar or slightly warmer than today on the basis of the record of warm water mollusks (e.g., [27, 28]).

MIS 5 in the substage e, is one of the most studied episodes, and best represented worldwide. It encompasses from ca. 130 ± 2 to 119 ± 2 ka [29] and the SST was approximately 2°C higher than the present one (e.g., [30, 31]). Comparing with other interglacials, MIS 5e has the best records of SST [18]. Evidence of warm water benthic mollusks, and changes in their geographic distribution was found in MIS 5e (e.g., [5, 26, 27, 32–48]).

MIS 1 encompasses the last 11.7 ka [49], when the last glaciation is considered to be ended [50]. There is an increase of the SST and humidity worldwide (e.g., [4, 51–56]) with some records in the Southern Hemisphere (e.g., [37, 49, 50, 57–59]). This phenomenon is reflected worldwide in the biotic communities with changes in composition, abundance, diversity and distribution (e.g., [60–62]).

In the Argentine Patagonian region, the stages MIS 11 to MIS 1 are represented, as well as ingressions older then MIS 11, but with poor records and fossil content [32].

2. Area of study

2.1 Quaternary marine deposits from the northern argentine Patagonia

Along the Argentine coast, broad extensions of the littoral of the Buenos Aires Province (BAP) and north Patagonia were affected by accumulation and erosion processes produced by sea level oscillations during Quaternary transgressions (e.g., [63, 64]).

Northern Patagonia has been divided into two regions according to their province. The first groups three areas of the south of the BAP (A-C) and record the presence of the interglacials ≥ MIS 9, MIS 5e and MIS 1 [5, 65, 66], and the other one belongs to region D, and records the presence of the interglacials ≥ MIS 9, MIS 7, MIS 5e, and MIS 1 (**Figure 1**).

In region A, from Peninsula Verde to Otero Island, the transgressive deposits have been assigned to the Sangamon (? Late Pleistocene), which are represented in the area of the Colorado River delta by paleocliffs associated with coast lines up to 10 m height [67]. These marine deposits are assigned to the oldest interglacials because of their geomorphological, altimetric and cementation similarity [68]. Among them, there are scarce, thin, and isolated deposits on the continent which are assigned to ≥ MIS 9 [68, 69].

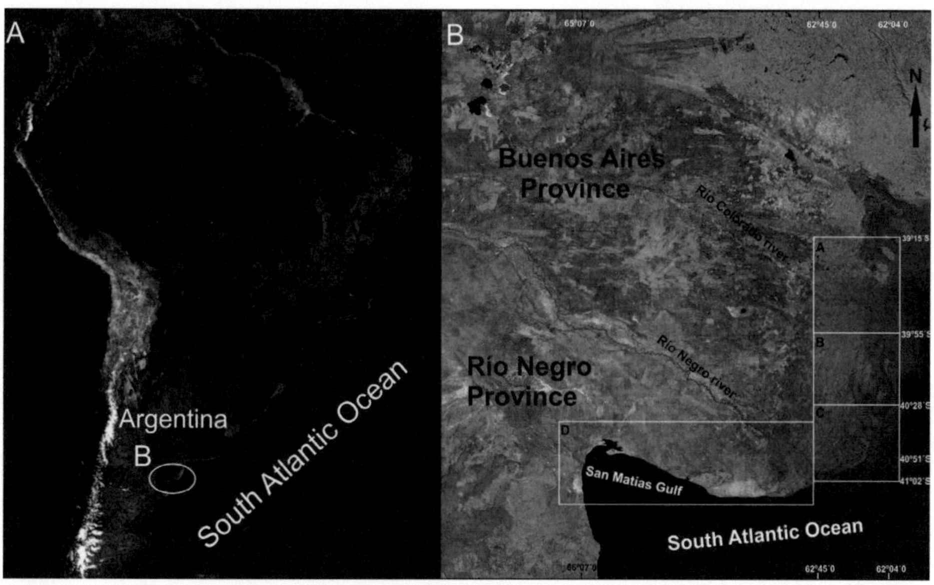

Figure 1.
(A) Map of study, and (B) four areas in the northern Patagonia in Argentina.

The terminal area of the delta is formed by the marine deposits of the MIS 1 ingression, which are beach ridges or intertidal environments [68].

In region B, which extends from the Otero Island to near the Jabalí Island, Weiler [70] correlated the Pleistocene and Holocene marine deposits of the central area with three transgressive events of the Late Pleistocene and Holocene, and related them to environments of barriers and coastal lagoons of the transgression of the Sangamon interglacial (with a minimum age of 43 Ka), the interstadial transgression of the middle Wisconsin (38.5 to 25 Ka), and the postglacial transgression of the MIS 1 (middle Holocene, between 5 and 5.2 Ka). More recently, Schnack et al. [71] related the oldest deposits of the area described by Weiler [43] with the Interglacial MIS 5e, considering the radiocarbon datings as minimum ages.

In region C, from the Jabalí Island up to Villa 7 de Marzo, Fucks et al. [68] reinterpreted the stratigraphic sequences, assigning a minimum of four transgressive cycles. Beach ridges, as well as beach strand plains and tidal plains, with maximum altitudes of 6 m a.s.l. and very clear morphologies are present from the coast to the present day continent, particularly in Isla Jabalí. Above them, at altitudes of 8 to 10 m a.s.l. that increase gradually to over 30 m a.s.l., clear ridges could be probably related to MIS 5e. These could have been originated in two ≥ 9 transgressive events.

In region D, north of San Matías Gulf from near the El Cóndor beach up to Las Grutas beach, there are records of the interglacials MIS 7, MIS 5e and MIS 1 plus a fourth one, 60 m height, that probably corresponds to an interglacial ≥ MIS 9. According to Fucks et al. [72] the deposits of interglacials MIS 7 and MIS 5e correspond to Baliza San Matias and San Antonio formations, respectively, and those of Holocene age have no designation.

The main geomorphological features are littoral ridges formed by high energy conditions, although, deposits corresponding to intertidal environments, coastal lagoons, spit sand cliffs forms have been described mainly for the Holocene transgressive event as well (e.g.,[5, 68, 72–75]).

All these deposits contain marine mollusks, particularly gastropods and bivalves (**Figure 2**) (**Table 1**) [5, 35–37, 65, 82].

Figure 2.
(A) Profile of Holocene deposit with Tagelus plebeius *in life position (region a, MIS 1); (B) Holocene littoral ridge in canal Villalonga (region B, MIS 1); (C) Pleistocene outcrops near Cardenal Cagliero locality (region C, MIS 9) and (D) Holocene cliff near San Antonio Este Harbor (region D, MIS 1).*

Province	Área	Coordinates (Lat-long)	Sites	Ages (14C and ESR)	Altitude (m.a.m.s.l)	Cites
South of Buenos Aires	Delta del río Colorado	39°20′S;62°04′W–39°55′S;62°08′W	P. Verde (39°21′S;62°5.9′W)	2170 ± 86 ka	5–2.5	[76]
			Pta Laberinto – rio Colorado Viejo (39°30′S–39°50′S)	6.63 ± 0.12 ka–0.409 ± 0.10 ka	10–7.5 7.5 and 2	[74]
			Sur del río Colorado viejo (39°53′S; 62°10′W)	9.46 ± 0.12 ka–0.407 ± 0.10 ka	5–2.5	[74]
	Bahía Anegada	39°55′S;62°08′W–40°28′S;62°11′W	Pleistocene deposits (40°03′S–40°26′S)	43 ka, 38.55 ka, 25 ka	40–25	[70, 77]
				43 ka, 38.8 ka, 31 ka	10–7	[75]
			Canal Villalonga (40°1′S; 62°19′W)	5.98 ± 0.10 ka–3.69 ± 0.10 ka	4–2.5	
			Los Pocitos (42°25′S; 62°25′W)	4.4 ± 0.08 and 4.5 ± 0.09 ka	3	[72, 74, 78]
	Isla Jabalí – Villa 7 de marzo	40°36′S;62°11′W–41°1′S;62°45′W	Oeste de isla Jabalí (40°40′S;62°30′W)	30.78 ± 1.65 ka–28.4 ± 0.80 ka	3–9.5	[75, 79]
			Isla Jabalí (40°34′S;62°13′W)	5. 37 ± 0.11 ka-2.17 ± 0.11	4–5	[37]
			Faro Segunda Barranca (40°46′S;62°16′W)	102–108 ka 94.5, 79 and 72.7 ka 28 a 40 ka	8–10	[80, 81]
Northern of Rio Negro	Norte del golfo San Matías	40°51′S;65°7′W–41°02′S; 62°49′W	Caleta Falsa correlation with localities: south of Piedras Coloradas	≥230 and ≥169 ka	8–12	[80, 81]
			Baliza San Matias (40°42′S; 64°51′W)	97.3–83 ka	8	[73]
			La Rinconada (40°41′S;65°9′W)	107–91 ka	10	
			Las Grutas (40°48′S;65°4′W)	70.3–66.8 ka	10	
			La Conchilla (40°49′S/64°52′O)	2.43 ± 0.60 ka	1.50	[5]

Table 1.
Absolute dating available from the study área.

3. Materials and methods

3.1 Methodology for marine mollusks associated with littoral deposits

A total of 84 localities were studied in two areas, 31 Pleistocene, 29 Holocene, and 24 modern ones. In these localities 7385 valves and shells of mollusks were collected. Each coastal deposit to be studied is identified on topographic maps. At each level of the site, a volumetric sample of 1 dm^3. In contrast, at modern beach sites the sample is taken in a quadrant 1 m x 1 m along transects perpendicular to the coast line.

Each fraction of biogenic content recovered from the sieves (2.80, 1.40 and 0.080 mm) was identified, measured with digital caliper, and labeled. Species were identified through catalogs and specific literature.

Valves and shells found in the marine deposits are considered as assemblages representing the accumulation of non-contemporary individuals in a single set, and occurs because the time of generation of these individuals is faster than the burial rates. In this context, it has to be taken into account the changes produced during the transition of the animal remains from the biosphere to the lithospere, which is studied through a discipline called taphonomy (etymologically derived from the Greek taphos, tomb, and nom, law). Taphonomy was defined by Efremov [83] as the science that studies the laws of burial, and accepted that taphonomic processes lead to the loss of information and are the cause of gaps in the fossil record. Currently this concept has been reversed since numerous studies (e.g., [84–87]) support the idea that the faunal associations of both current and fossil valves provide relevant information on living communities, or paleocommunities, being able in both cases to reconstruct the environments or paleoenvironments from the analysis of the faunal associations and thus interpret environmental and climatic changes.

The studied mollusks form transported faunal associations, which, according to different authors, preserve compositional fidelity concerning taxonomy and relative abundance of the communities that inhabited each environment within the considered period, and represent the accumulations of non-contemporary individuals as a whole.

4. Results

4.1 Marine malacology of northern Patagonia

The four areas identified in northern Patagonian according to their geomorphology (A-D), in which 84 sites were studied, yielded 78 species (42 bivalves and 36 gastropods). Eleven of them are micromollusks: *Heleobia australis*, *Olivella tehuelcha*, *Parvanachis isabellei*, *Turbonilla argentina*, *Turbonilla paralaminata*, *Chrysallida multituberculata* (gastropods) and *Nucula nucleus*, *Ennucula grayi*, *Carditamera plata*, *Corbula patagonica*, and *Corbula lyoni* (bivalves) (**Tables 2** and **3**).

4.2 Malacological analysis

Seventy species were identified in areas A-C (37 bivalves and 33 gastropods), and 45 species in region D (19 bivalves and 24 gastropods), with a similarity of 51.3% in bivalve species and 48.5% in gastropods.

In areas A-C, all the studied sites assigned to the interglacial $\geq$ MIS 9, are paleobeaches and littoral ridges; i.e., high energy environments in which the marine fauna is euryhaline (salinity > 30–35 gr/l) and of sandy substrate. Warm water species prevailed in bivalve associations (50–67%) compared to the other interglacials recorded in the area. However, no bivalve or gastropod of warm lineage that constitutes itself a paleoindicator was found.

Bivalves	Salinity	Life habit	Depth (m)	Substrate	Trophic type	Distribution area
Nucula (N.) nucleus [88]	E	I	0–200	S	D	23°S–53.5°S
Ennucula grayi (d'Orbigny, 1846)[1]	E	I	5–1850	S	D	22.93°S–55.5°S
Adrana electa [89]	E	I	20–75	S	D	22.93°S–39°S*
Glycymeris (G.) longior [90]	E	I	10–75	S	Sf	10°S–42°S
Mytilus edulis platensis (d'Orbigny, 1846)[1]	P–E	Ep	0–50	H	Sf	68°N–55.5°S
Brachidontes (B.) rodriguezii (d'Orbigny, 1846)[1]	P–E	Ep	0–25	H	Sf	34°S–42°S
Aulacomya atra [91]	E	Ep	0–30	H	Sf	34°S–55.5°S
Aequipecten tehuelchus (d'Orbigny, 1842)[1]	E	Ep	10–120	M	Sf	21°S–42.58°S
Atrina seminuda [92]	P–E	Ce	0–3	H	Sf	35°N–35°S*
Plicatula gibbosa [93]	E	Ce	0–120	H	Sf	35.3°N–34°S*
Ostreola equestris [94]	P–E	Ce	0–80	H	C	37°N–42°S
Ostrea puelchana (d'Orbigny, 1841)[1]	P–E	Ce	0–70	H	C	22°S–42°S
Crassostrea rhizophorae [95]	P–E	Ce	0–50	H	C	21.4°S–35°S
Crassostrea gigas [96]	E	Ce	0–40	H	Sf	Cosmopolitan
Diplodonta (D.) patagonica (d'Orbigny, 1842)[1]	E	I	36–102	S	Sf	21°S–42.58°S
Diplodonta (F.) vilardeboana (d'Orbigny, 1846)[1]	E	I	25–77	S	Sf	21°S–42°S
Carditamera plata [97]	E	I	17–70	S	Sf	23°S–39°S*
Trachycardium muricatum [88]	E	I	0–11	S	Sf	35°N–42°S
Mactra guidoi [98]	P–E	I	0–25	S	Sf	34°S–42°S
Mactra isabelleana (d'Orbigny, 1846)[1]	P–E	I	0–25	S	Sf	23°S–42°S
Raeta (R.) plicatella [99]	P–E	I	0–11	S	Sf	39°N–41°S
Mesodesma mactroides [100]	E	I	0–20	S	Sf	23°S–41°S
Solen tehuelchus [101]	E	I	10–18	S	Sf	23°S–39°S*
Macoma (P.) uruguayensis [102]	E	I	18–70	S	D	29°S–39°S*
Angulus gibber [97]	E	I	13–55	S	D	23°S–43°S
Abra (A.) aequalis [94][1]	E	I	0–50	S	D	35°N–23°S*
Tagelus (T.) plebeius [103]	P	I	0–10	S	Sf	42°N–54°S
Tivela isabelleana (d'Orbigny, 1846)[1]	E	I	0–55	S	Sf	21°S–42°S
Anomalocardia brasiliana [104]	P–E	I	0.3–5	S	Sf	18°N–39°S*
Pitar (P.) rostratus (Philippi, 1844)[1]	E	I	10–100	S	Sf	22°S–38.7°S*
Amiantis purpurata (Dillwyn, 1817)[1]	E	I	0–20	S	Sf	19°S–43°S
Retrotapes exalbidus [105]	E	I	50–70	S	Sf	34°S–55.5°S

Bivalves	Salinity	Life habit	Depth (m)	Substrate	Trophic type	Distribution area
Ameghinomya antiqua [106]	E	I	5–50	S	Sf	34°S–54°S
Panopea abbreviata (Valenciennes, 1839)[1]	E	I	25–75	S	Sf	23°S–48°S
Corbula (C.) patagonica (d'Orbigny, 1846)[1]	E	I	15–90	S	Sf	23°S–43°S
Corbula (C.) lyoni [107]	E	I	11–67	S	Sf	19°S–43°S
Cyrtopleura (S.) lanceolata (d'Orbigny, 1846)[1]	E	I	10–27	S	Sf	6°S–42°S
Barnea lamellosa (d'Orbigny, 1846)[1]	E	I	15–150	R	Sf	34°S–43°S
Lyonsia (L.) alvarezii (d'Orbigny, 1846)[1]	E	I	50–86	S	Sf	38.3°S–41°S
Periploma ovatum (d'Orbigny, 1846)[1]	E	I	?	S	Sf	35°S–40.5°S
Thracia similis [108]	E	I	50–86	S	Sf	22°S–42.58°S

Ep, epifaunal; I, infaunal; Ce, cemented; H, hard; S, soft; C, carnivorous; D, detritivorous; He, herbivore; Sf, suspension feeder; O, oligohaline (3–8‰); M, mesohaline (8–18‰); P, polyhaline (18–30‰); E, euhaline (>30–35‰).
[*]Taxa found in the studied area.
[1]The references of species are found on MolluscanBase eds (2021).

Table 2.
Ecological requeriments and distribution of bivalves.

Interglacial MIS 5e is represented in all the studied sites mostly by littoral ridges in which most associations are euryhaline, of sandy substrate and subordinate rocky substrate. Most species are epifaunal except in region A, prevailing filter feeders and carnivores. The proportion of warm water species in this interglacial is lower than in the previous one (44–50%) being outstanding the record of the warm lineage bivalve *Crassostrea rhizophorae*, excellent paleoindicator. In region D, this proportion is only 27% of the bivalves.

Interglacial MIS 1 was recorded in the whole study area, with two types of deposits in areas A-C: littoral ridges (high energy environments) and tidal plains (low energy environments). In the first ones, the malacofaunal associations are mostly euryhaline, of sandy substrates. In the second ones instead, the associations vary in salinity from oligohaline to mesohaline-polyhaline (salinity between 3 and 30 gr/l), of fine sand substrate, mostly epifaunal and filter feeders prevailing infaunal and carnivores in region A. This latter would be related to the modern geomorphological features of low energy environments (wide tidal plains, tidal channels and non-functional fluvial courses), that resulted in the formation of islands which can be seen in all the southern coast of the Buenos Aires Province. This interglacial MIS 1 is recognized in area D in littoral ridges, being the malacofaunal associations mostly euryhaline, of sandy substrate with rocky subordinate. In the associations of MIS 1 there is 45 to 50% of warm water species in areas A-C, unlike areas D in which this proportion is only 18% of the total bivalves.

In modern beaches of the south of BAP, but not in the northern sector of Bahía Anegada there are sandy beaches together with mud-sandy ones, and malacofaunal associations correspond to marine parameters of high energy, euryhaline of sandy substrate with scattered rocky substrate, mainly in area C. Instead, the modern

Gastropods	Salinity	Life habit	Depth (m)	Substrate	Trophic type	Distribution area
Nacella (P.) magallanica [104]	E	Ep	0–200	H	He	38.5°S–55.5°S
Diodora (D.) patagonica (d'Orbigny, 1841)[1]	E	Ep	0–15	H	He	11°N–45°S
Fissurella radiosa radiosa (Lesson, 1831)[1]	E	Ep	0	H	He	48°S–55°S
Lucapinella henseli [109]	E	Ep	0–55	H	He	23°S–
Calliostoma carcellesi (Clench and Aguacho, 1940)[1]	E	Ep	0–60	S	He	40.37°S–41.67°S
Calliostoma coppingeri [110]	E	Ep	13–86	S	He	30°S–44.21°S
Tegula (A.) patagonica (d'Orbigny, 1835)[1]	E	Ep	0–57	H	He	23°S–54°S
Tegula atra (Lesson, 1830)[1]	E	Ep	0–9	H	He	38°S–55°S
Bostrycapulus odites [111]	E	Ep	0–46	H	Sf	25°S–45.8°S
Crepidula argentina [112]	E	Ep	30–50	H	Sf	38°S–41.03°S
Crepidula dilatata [113]	E	Ep	0–66	H	Sf	35°S–55.8°S
Notocochlis isabelleana (d'Orbigny, 1840)[1]	E	I	0–113	S	C	22.4°S–42.58°S
Heleobia australis (d'Orbigny, 1835)[1]	O, P, M	Ep	0–60	M	He	24°S–41°S
Epitonium (E.) georgettinum (Kiener, 1838)[1]	E	Ep	0–101	M	He	23.37°S–44.27°S
Epitenium striatellum (Nyst, 1871)[1]	E	Ep	30	M	He	23°S–41°S
Trophon patagonicus (d'Orbigny, 1839)[1]	E	Ep	0–50	H	C	32°S–40°S
Trophon geversianus [114]	E	Ep	0–58	H	C	36.42°S–54.98°S
Urosalpinx cala [107]	E	Ep	28–28	H	C	32°S–41°S
Zidona dufresnei [115]	E	Ep	10–90	S	C	23°S–42°S
Adelomelon (P.) brasiliana [116]	E	Ep	0–250	S	C	23°S–52°S
Adelomelon beckii [117]	E	Ep	40–75	S	C	20° S–52°S
Odontocymbiola magallanica [104]	E	Ep	10–200	M	C	35°S–55.2°S
Marginela martini (Petit, 1853)[1]	E	Ep	10–80	S	C	22.93°S–42°S
Olivella (O.) tehuelcha [118]	E	Ep	15–57	S	C	23.69°S–43°S
Olivancillaria urceus (Röding, 1798)[1]	E	Ep	5–50	S	C	19°S–42°S
Olivancillaria carcellesi [119]	E	Ep	0–22	S	C	23°S–42.5°S
Olivancillaria uretai [119]	E	Ep	0–30	S	C	23°S–40.6°S
Buccinanops monilifer [120]	E	Ep	0–50	S	C	35°N–42°S

Gastropods	Salinity	Life habit	Depth (m)	Substrate	Trophic type	Distribution area
Buccinanops cochlidium [105]	E	Ep	5–66	S	C	23°S–42.58°S
Buccinanops globulosus [120]	E	Ep	0–6	S	C	35°S–46°S
Buccinanops uruguayensis [107]	E	Ep	15–45	S	C	24°S–42°S
Parvanachis isabellei (d'Orbigny, 1839)[1]	E	Ep	10–65	S	C	30°S–54°S
Costoanachis sertulariarum (d'Orbigny, 1839)[1]	E	Ec	0–20	S	C	35°N–54°S
Turbonilla argentina [121]	E	Ec	18–57	S	C	35°S–41°S
Turbonilla paralaminata [122]	E	Ec	30–65	S	C	39°S–41°S
Chrysallida multituberculata [122]	E	Ec	30–65	S	C	40°S–46°S
Siphonaria lessoni [123]	E	Ep	0	H	He	32°S–55.22°S

Ep, epifaunal; I, infaunal; Ce, cemented; Ec, ectoparasite; H, hard; S, soft; M, mixed; C, carnivorous; D, detritivorous; He, herbivore; Sf, suspension feeder; O, oligohaline (3–8‰); M, mesohaline (8–18‰); P, polyhaline (18–30‰); E, euhaline (>30–35‰).
[1]*The references of species are found on MolluscanBase eds(2021).*

Table 3.
Ecological requeriments and distribution of gastropods.

beaches of area D are larger often exceeding hundreds of meters wide. There are two types of beaches regarding the granulometry: a low intertidal sector of fine to medium sand with high distal sectors of gravels, organogenic in composition, and a low intertidal sector and high distal one of fine-medium sand. Both are associated with high energy environments where the malacological associations are mostly euryhaline, of sandy substrate, with less proportion of fauna of scattered rocky substrate. There is one exception, the modern beach of Villa 7 de Marzo which has a particular feature, fine sand substrate with abundance of two bivalves *Plicatula gibbosa* and *Ostrea puelchana*, and the gastropod *Crepidula*, which could be due to the influence of the Negro River. This is also observed in the southern area of this river, in El Cóndor Beach, which has a similar environment and the same mollusks. The modern malacofaunal associations bear 33 to 50% of warm water species in regions A-C, and 31% in area D.

In area D, all the analyzed sites of MIS 7 correspond to paleobeaches. They are currently represented by coastal platforms, of high energy, with mostly euryhaline malacofaunal associations of sandy substrate and subordinated rocky substrate. Respect to the indicators of sea water temperature, these associations are formed only by 20% of warm water bivalves, although it is recorded the gastropod *Tegula atra*, probably indicator of warm water.

In MIS 5e there is a slightly higher proportion of associations of warm waters (27%) with respect to the previous interglacial, being conspicuous the presence of *Anomalocardia brasiliana* and *Tegula atra*, the first one an excellent paleoindicator, and quite probably also the second one.

In MIS 1 the associations of warm water are in lesser proportion than those of MIS 5e (18%), being outstanding the record of *Mesodesma mactroides*. The bivalves *Aulacomya atra*, *Retrotapes exalbidus*, *Ameghinomya antique* and the gastropods *Fisurella radiosa radiosa* and *Crepidula dilatata* are recorded in the beaches of Río

Negro Province. These species have not been described in the modern marine malacofauna of areas A-C (**Figures 3–6**).

4.3 Warm water bivalves and gastropods

4.3.1 Interglacials ≥ MIS 9 and MIS 7

These two interglacials recorded in area D were not recorded in the northeast of BAP. In areas A-C, a total of nine species of mollusks were recorded in the

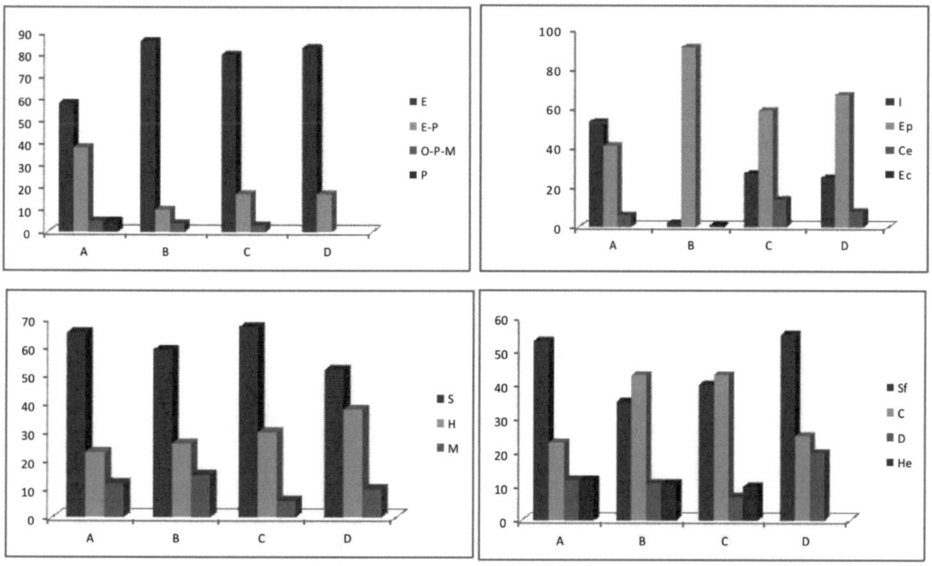

Figure 3.
Paleoecological features of all regions (A–D) in interglacial MIS 5e.

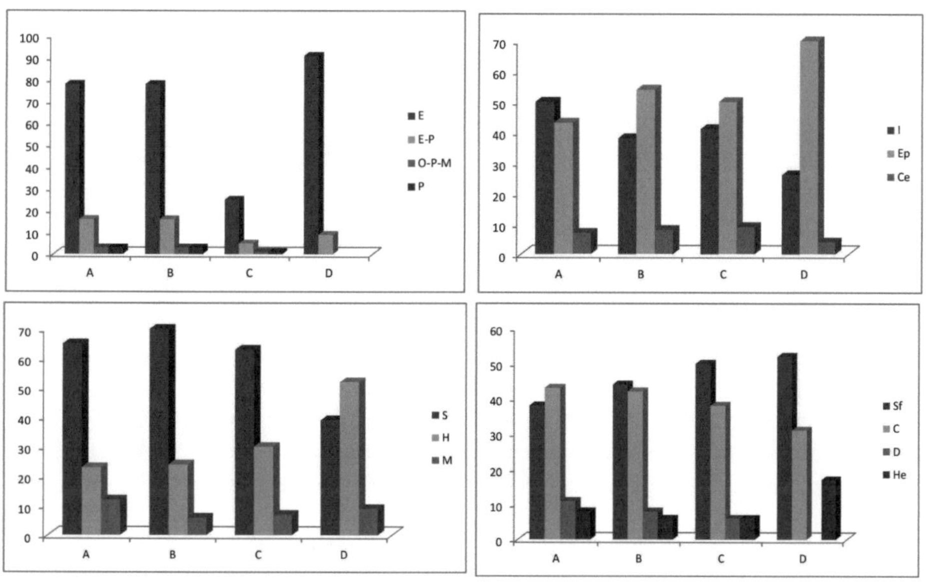

Figure 4.
Paleoecological features of all regions (A–D) in interglacial MIS 1.

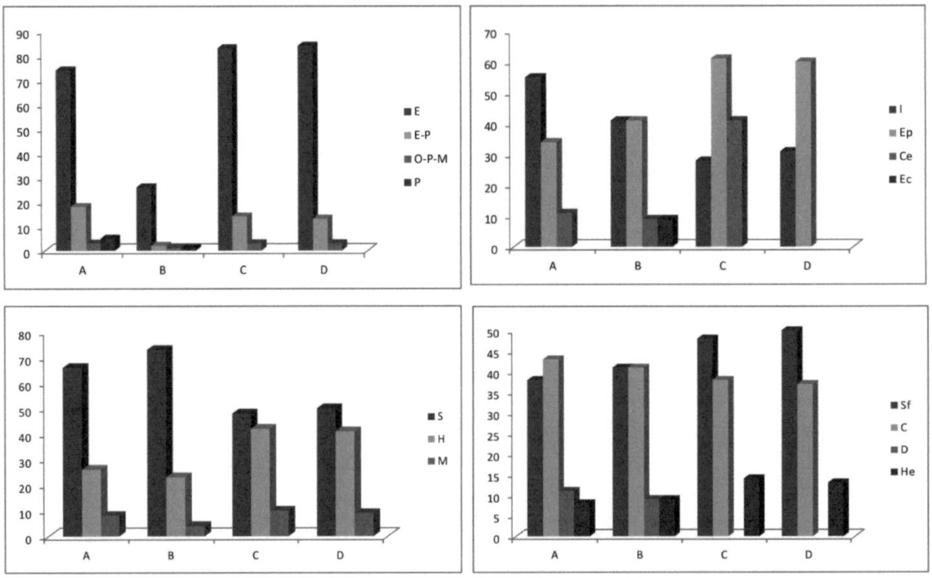

Figure 5.
Paleoecological features of all regions (A–D) in modern beaches.

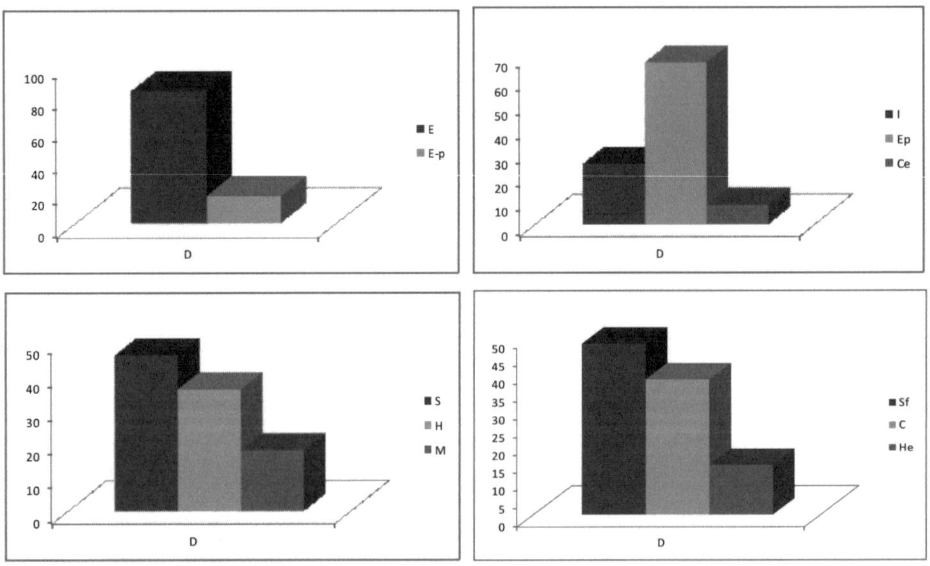

Figure 6.
Paleoecological features of region D in interglacial MIS 7.

Interglacial ≥ MIS 9, and among them, there is 50–67% of warm water species of bivalves, being this the oldest record of marine mollusks for the BAP. In MIS 7 analyzed in area D, a total of 11 species was recorded with 20% of warm water bivalves. Most recorded species in both interglacials still inhabit the modern coasts of Argentina, except for the gastropod *Tegula atra*, absent since the MIS 1.

Both interglacials (≥MIS 9 and MIS 7) revealed the presence of warm water mollusks which are not recorded in the marine deposits of the northeast of BAP where these deposits have not been preserved [124]. The record of these interglacials in the BAP is a novelty in the analysis of gastropods and bivalves. Whereas Aguirre et al. [27, 28] reported that in the coasts of southern Patagonia, in areas such as Bahía Vera-Camarones (44.2° to 45°S) and Bahía Bustamente-Caleta Olivia (44.9°–45.3°S, Chubut Province, Argentina), the environmental conditions (substrate, depth, and energy conditions) during the late Pleistocene (MIS 7 and MIS 5e) suggest SST similar to those of the modern littoral and even slightly higher than present, recording faunas of warm to temperate waters.

4.3.2 Interglacial MIS 5e

In the marine deposits of this interglacial in area D, 44 molluskan species (25 bivalves and 19 gastropods) were recorded. These deposits are represented by littoral ridges and tidal plains along the south of the BAP coast, and by littoral ridges along area D, with scarce content of calcium carbonate, favoring the record of mollusks. In area C the associations of MIS 5e and MIS 1 have respectively 34 and 33 species. This similarity was also found in area D, in which 22 species were recognized for MIS 5e and 23 for MIS 1. Unlike area D, the mollusks of MIS 5e of the northeast of the BAP are characterized by less abundance and diversity of species related to those of the Holocene MIS 1 [32]. This could be due to a less representation of the Interglacial MIS 5e, and because most valves and shells of Pleistocene deposits are dissolved and/or crystallized, preventing the species identification.

Between 50 and 44% of warm water species of bivalves recorded in the Interglacial MIS 5e of area D, are represented in areas A-C, whereas this relationship is only 27% of the species from area D. Warm water mollusks were recognized in the study area, among the most prominent species of this interglacial, in areas A-C, is the bivalve *Crassostrea rhizophorae* (**Figure 7A**). This is a warm lineage species that inhabits currently the Caribbean, Venezuela, Surinam, and Brazil up to Uruguay, but is not recorded in the present Argentine coast. However, it is recorded as fossil in the northeast and south of the BAP in the MIS 5e (e.g., [33, 125]).

Another warm lineage species found in area D, is the bivalve *Anomalocardia brasiliana*, recorded in the area north of the Negro River (**Figure 7B**). This is the most austral record of the species in the Argentine coast. This species is distributed currently from the French Antilles (18°N) up to the coasts of Brazil (33°S), being an infaunal surface species able to support wide salinity ranges (e.g., [126, 127]).

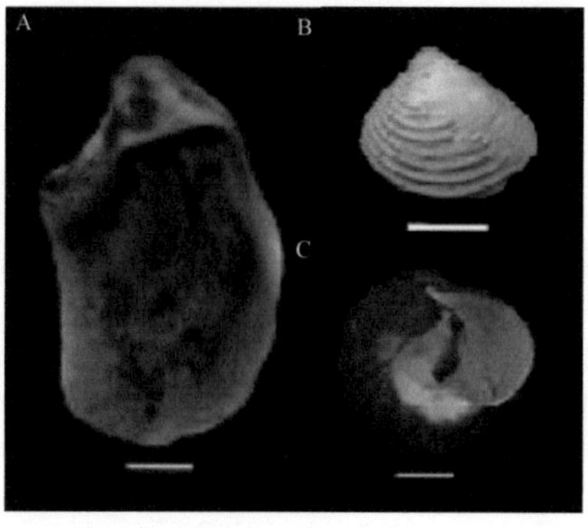

Figure 7.
The warm species of Pleistocene. (A) Crassostrea rhizophorae *(Guilding) (MLP: 34.012, MIS 5e, region B), (B)* Anomalocardia brasiliana *(Gmelin) (CEGH-UNC: 25.609. MIS 5e, region D) and (C) Tegula atra (lesson) (CEGH-UNC: 25.615, MIS 5e, region D).*

As fossil, it was found in Uruguay, both in the MIS 5e of the Nueva Palmira Formation [43] and in the Holocene of the Villa Soriano Formation [128]. In Argentina, this species is reported for marine deposits within the Pampiano Formation of Lomas de Zamora (34°46′S, northeast of BAP; [129]), as well as in the localities of Magdalena, Punta Piedras (BAP) and south of Entre Ríos Province [124]. This species was found in the Pleistocene deposits of Bahía Blanca (south of BAP) together with *Crassostrea rhizophorae* [33].

Tegula atra is distributed in intertidal and subtidal shallow rocky substrates, up to 6 m depth (**Figure 7C**) [130]. It is among the most abundant species of the Pleistocene of area D, and is well preserved in deposits of MIS 7 and MIS 5e [5] but it is not recorded in deposits of MIS 1. Whereas in the Pacific coast, *Tegula atra* is recorded in the late Pleistocene and Holocene deposits of the northern and southern coast of Peru (e.g., [131, 132]). In Chile, it is recorded in late Pleistocene deposits of Caleta Coloso (23°45′S/70°28′W), north and south of Antofagasta (23°37’ S) [45], as well as in archeological sites such as the late Pleistocene-middle Holocene of Quebrada de Lazareto (south of Chile) [133], middle and late Holocene of the IV Region, Los Vilos (e.g., [134]) and Holocene deposits of the Magellan Straight [135]. *Tegula atra* is currently distributed in the Pacific coasts from Pacasmayo (7°24′S, Peru) up to the Magellan Straight (53°S) [136], but there is no evidence of living specimens in the south Atlantic coasts.

4.3.3 Interglacial MIS 1

A total of 58 species (31 bivalves and 27 gastropods) was recorded in the marine malacofauna of the Interglacial MIS 1 of northern Patagonia (areas A-D), which differs from the northeast of BAP where Aguirre [89] reported a total of 62 species (25 bivalves and 37 gastropods). Concerning the molluscan composition, in the northeast of BAP, gastropods are more abundant than bivalves both in number of species and of individuals. As a comparison, among the regions studied, area A recorded 51 species (29 bivalves and 22 gastropods), and area B 49 species (25 bivalves and 24 gastropods), being in both regions, bivalves more numerous than gastropods. Whereas in area C, 34 species (17 bivalves and 17 gastropods) were recorded, unlike area D where 42 species (20 bivalves and 22 gastropods) were recorded, being in this latter the number of gastropods slightly higher than bivalves.

The marine deposits of MIS 1 in areas A-C are formed by tidal plains and littoral ridges. Tidal plains yielded mainly *Tagelus plebeius* among bivalves and *Heleobia australis* among gastropods. These species support variable salinity, being recorded in oligohaline–mesohaline–polyhaline associations (salinity between 3 and 30 gr/l), typical of low energy environments, most of them with low diversity indexes. In the littoral ridges, the diversity indexes are mostly higher than those of tidal plains. They yielded among bivalves *Pitar rostratus*, *Amiantis purpurata*, *Ostreola equestris*, and among gastropods *Buccinanops cochlidium* and *Heleobia australis*, typical of high energy environments.

The marine deposits and their malacofauna in the northeast and south (areas A-C) of the BAP are similar in the two types of deposits of the Interglacial MIS 1. The tidal plains are represented in regions A-C and are related to the development of the Colorado River and Bahía Anegada. The most common species are *Heleobia australis*, *Tagelus plebeius* (in life position) and *Corbula patagonica*, being similar to the marine malacofauna of the northeast of the BAP of the Canal 18 Member of the Las Escobas Formation (Holocene, MIS 1) and the estuarine facies of the Mar Chiquita Formation (**Figure 8**).

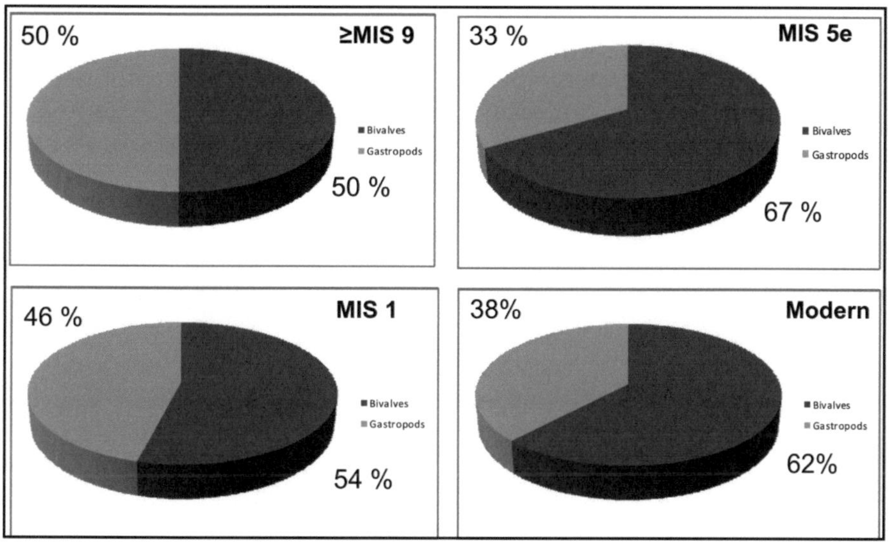

Figure 8.
Warm water vs. cold water during the quaternary in all regions (A–D).

5. Final comments

In the last 400.000 years there were variations in the molluskan paleocommunities of north Patagonia (southern BAP and north of Río Negro Province). The marine malacofauna of this area is composed by two associations, the first one is formed by Holocene sites with abundance of *Heleobia australis* and *Corbula patagonica*, and the other, mainly by Pleistocene sites and modern beaches with *Amiantis purpurata* as the most abundant, together with *Buccinanops globulosus*, *Bostrycapulus odites* and *Tegula patagonica*. The record of *Crassostrea rhizophorae*, *Anomalocardia brasiliana* and *Tegula atra*, in north Patagonia suggests that the interglacials MIS 7, MIS 5e and MIS 1 were slightly warmer than today. These latter species, except for *Tegula atra* live today in lower latitudes. However, the associations determined for the analyzed interglacials did not change concerning the faunal composition as a whole, and, except for the latitudinal changes of the three species mentioned above, and the record of temperate to cold water associations since the Interglacial MIS 1 in north Río Negro Province (ecotone), the composition remained similar, showing only changes in abundance of species.

Author details

M.P. Charó[1,2]

1 Servicio de Hidrografía Naval (CONICET), Caba, Argentina

2 Instituto Superior de Correlación Geológica (INSUGEO/CONICET-UNT), San Miguel de Tucumán, Argentina

*Address all correspondence to: charomelisa@gmail.com

References

[1] Rutherford, S. and D'Hondt, S., 2000. Early onset and tropical forcing of 100.000-year Pleistocene glacial cycles. Nature, 408, 72-75.

[2] Reyes, A. V.; Carlson, A. E., Beard, B. L., Hatfield, R. G., Stoner, J. S., Winsor, K., Elke, B. and Ullman, D. J. 2014. South Greenland ice-sheet collapse during Marine Isotope Stage 11. Nature, 510: 525-528.

[3] Raymo, M. E., Lisiecki, L. E., Nisancioglu, K. H. 2006. Plio-Pleistocene ice volume, Antarctic Climate, and the global d018 record. Science, 313, 492-495.

[4] Salvigsen, O.; Elgersma, A. and Landvik, J.Y. 1991. Radiocarbon dated raised beaches in northwestern Wedel Jarlsberg L and, Spitsbergen, Svalbard. Wyprawy Geograficzne na Spitsbergen, Lublin, Poland, 9-16.

[5] Charo, M. P., 2014. Caracterización paleoambiental y paleodiversidad malacológica en los depósitos marinos cuaternarios del norte patagonico (sur de Buenos Aires y Norte de Rio Negro). Facultad de Ciencias Naturales y Museo, La Plata, Argentina, 306.

[6] Emialiani, C., 1955. Pleistocene temperatures. Journal of Geology, 63, 538-578.

[7] De Abreu, L., Abrantes, F. F., Shackleton, N. J., Tzedakis, P. C., McManus, J. F., Oppo, D. W. and Hall, M. A., 2005. Ocean climate variability in the eastern North Atlantic during interglacial marine isotope stage 11: A partial analogue to the Holocene? Paleoceanography, 20. DOI:10.1029/2004PA001091.

[8] Loutre, M. F. and Berger, A., 2003. Marine isotope Stage 11 as an anologue for the present interglacial. Global Planet Change, 36: 209-217.

[9] Wu, N.; Chen, X., Rousseau, D. D., Li, F., Pei, Y. and Wu, B. 2007. Climatic conditions recorded by terrestrial mollusc assemblages in the Chinese Loess Plateau during marine Oxygen Isotope Stages 12-10. Quaternary Science Reviews, 26:1884-1896

[10] Ashton, N., Lewis, S. G., Parfitt, S. A., Penkman, K. E. M., Russel Coope, G. 2008. New evidence for complex climate Change in MIS 11 from Haxne, Suffolk, UK. Quaternary Science Reviews, 27: 652-668.

[11] Ortlieb L.; Guzman, N. and Marquardt, C. 2003. A Longer-Lasting and Warmer Interglacial Episode During Isotopic Stage 11: Marine Terrace Evidence in Tropical Western Americas. Earth's Climate and Orbital Eccentricity The Marine Isotope Stage 11 Question Geophysical Monograph, 137. DOI:10.1029/137GM12.

[12] Preece, R. C.; Parfitt, S. A., Bridgland, D. R., Lewis, S. G., Rowe, P. J., Atkinson, T. C., Candy, I., Debenham, N. C., Penkman, K. E. H., Rhodes, E. J., Schwenninger, J. L., Griffiths, H. I., Whittaker, J. E. and Gleed-Owen, C. 2007. Terrestrial environments during MIS 11: evidence from the Palaeolithic site at West Stow, Suffolk, UK. Quaternary Sciences review, 26 (9-10): 1236-1300. DOI:10.1016/j.quascirev.2006.11.016.

[13] Pedoja, K.; Regard, V., Husson, L., Martinod, J., Guillaume, B., Fucks, E., Iglesias, M. and Weill, P. 2011. Uplift of Quaternary shorelines in eastern Patagonia: Darwin revisited. Geomorphology, 127:121-142.

[14] Stirling, C. H.; Esat, T. M., Lambeck, K., Mc Culloch, M. T., Blake, S. G., Lee, D. C. and Halliday, A. N. 2001. Orbital Forcing of the Marine Isotope Stage 9 Interglacial. Science, 291 (5502): 290-293. DOI:10.1126/Science.291.5502.290.

[15] Zazo, C.; Goy, J.L., Dabrio, C.J., Bardají, T., Hillaire-Marcel, C., Ghaleb, B., González-Delgado, J.A. and Soler, V. 2003a. Pleistocene raised marine terraces of the Spanish Mediterranean and Atlantic coasts: records of coastal uplift, sea-level highstands and climate changes. Marine Geology, 194: 103-133.

[16] Green, C. P., Branch, N. P., Coope, G. R., Field, M. N., Keen, D. H., Wells, J. M., Schwenniger, J. L., Preece, R. C., Schereve, C. D., Canti, M. G. and Gleed-Owen, C. D., 2006. Marine Isotope Stage 9 environments of fluvial deposits al Hackney, north London, UK. Quaternary Science Reviews, 25 (1-2): 89-113.

[17] Dutton, A., Bard, E., Antonioli, F., Esat, T. M., Lambeck, K. and McCulloch, M. T. 2009. Phasing and amplitude of sea-level and climate change during the penultimate interglacial. Nature Geoscience, 2: 355-359. DOI:10.1038/ngeo470.

[18] Lang, N. and Wolff, E. W. 2011. Interglacial and glacial variability from the last 800 ka in marine, ice and terrestrial archives. Climate of the Past, 7: 361-380.

[19] Ortlieb, L. 1987. Neotectonic and Quaternary sea level variations in the gulf of California region. Bulletin of the Inqua Neotectonic Commission, 10: 28-31.

[20] Shackleton, N. J. 1987. Oxyigen isotopes, ice volume and sea level: Quaternary Science Review, 6: 183-190.

[21] Desprat, S., Sanchez Groñi, M. F., Turon, J-L., Duprat, J., Malaizé, B. and Peypouquet, J. -P. 2006. Climatic variability of marine isotope stage 7: direct land-sea-ice correlation from a multiproxy analysis of a north-western Iberian margin deep-sea core. Quaternary Science Reviews, 25 (9-10): 1010-1026.

[22] Isla, F. I. and Bujalevsky, G. 2008. Coastal Geology and morphology of Patagonia and Fueguian Archipelago. In: Rabassa, J. R. (ed) The Late Cenozoic of Patagonia and Tierra del Fuego., Elsevier Sciences Publication of Chile, 10: 227-240.

[23] Shackleton, N. J.; Berger, A. and Peltier, W. R. 1990. An alternative astronomical calibration of the lower Pleistocene timescale based on ODP Site 677. Earth and Environmental Science Transactions of the Royal Society of Edinburgh, 81 (4): 251-261.

[24] Hillaire-Marcel, C., Carro, C., Causse, O., Goy, J. L. and Zazo, C. 1986. Th/U dating of *Strombus bubonius*-bearing marine terraces in southeastern Spain. Geology, 14: 613-329.

[25] Zazo, C. and Goy, J. L. 1989. Sea level changes in the Iberian Peninsula during Last 200.000 years. Late Quaternary Sea-Level correlation and applications (D. B. Scott, P. A. Pirazolli and A. Honing, eds), NATO ASI Series C, 256, Blackwell Publication, 27-39.

[26] Zazo, C.; Goy, J. L., Hillaire-Marcel, C., Dabrio, C. J., Gonzalez Delgado, J. A., Cabero, A., Bardají, T., Ghaleb, B. and Soler, V. 2010. Sea level changes during the last and present interglacial in SalIsland (Cape Verde archipelago). Global and Planetary Changes. DOI: 10.1016/j.gloplacha.2010.01.006.

[27] Aguirre, M. L., Negro Sirch, Y. and Richiano, S. 2005. Late Quaternary molluscan assemblages from the coastal area of Bahía Bustamante (Patagonia, Argentina): Paleoecology and paleoenviroments. Journal of South Earth Sciences, 20: 13-32.

[28] Aguirre, M. L., Richiano, S. and Negro Sirch, Y. 2007. Moluscos de terrazas marinas cuaternarias del área de Camarones, Patagonia. Bolletín de la Societat d'Història Natural de las Balears, 14: 81-120.

[29] Hearty, P. J., Hollin, J. T., Neumann, A. C., O'Leary, M. J. and McCulloc, M. 2007.

Global sea-level fluctuation during the last Interglaciation (MIS5e). Quaternary Science Reviews, 26: 2090-2112.

[30] Murray-Wallace, C. V.; Beu, A. G., Kendric, G. W., Brown, L. J., Belperio, A. P. and Sherwood, J. E. 2000. Palaeoclimatic implication of the occurrence of the arcoid bivalve *Anadara trapezia* (Deshayes) in the Quaternary of Australasia. Quaternary Science Review, 19: 559-590.

[31] Rohling, E. J.; Grant, K., Hemleben, C. H., Siddall, M., Hoogakker, B. A. A., Bolshow, M., and Kucera, M. 2008. High rates of sea-level rise during the last interglacial period. Nature Geoscience, 1: 38-42.

[32] Aguirre, M. L., Donato, M., Richiano, S. and Farinati, E. A. 2011. Pleistocene and Holocene interglacial molluscan assemblages from Patagonian and Bonaerensian littoral (Argentina, SW Atlantic): Palaeobiodiversity and palaeobiogeography. Palaeogeography, Palaeoclimatology, Palaeoecology, 308: 277-292.

[33] Chaar, E. and Farinati, E. 1988. Evidencias paleontológicas y sedimentológicas de un nivel marino pleistocénico en Bahía Blanca, provincia de Buenos Aires, Argentina. Segundas Jornadas Geológicas Bonaerenses, Bahía Blanca, Argentina. Acta, 1: 47-54.

[34] Cohen, A. L., Parkington, J. E., Brundrit, G. B. and Van der Merwe, N. J. 1992. A Holocene marine climate record in mollusc shells from the Southwest African coast. Quaternary Research, 38 (3): 379-385.

[35] Charó, M. P., Fucks, E. E. and Gordillo, S. 2013b. Moluscos bentónicos marinos del Cuaternario de Bahía Anegada (Sur de Buenos Aires, Argentina): variaciones faunísticas en el Pleistoceno Tardío y Holoceno. Revista Mexicana de Ciencias Geológicas, V (30): 404-416.

[36] Charó, M. P., Gordillo, S., Fucks E. E. and Giaconi, L. M. 2014. Late Quaternary molluscs from the Northern San Matías Gulf (Northern Patagonia, Argentina), Southwestern Atlantic: Faunistic changes paleoenvironmental interpretation. Quaternary International. DOI:10.1016/j.quaint.2013.12.044

[37] Charó, M. P., Gordillo, S. and Fucks, E. E., 2013a. Paleoecology significance of Late Quaternary molluscan faunas of the Bahia San Blas area, Argentina. Quaternary International, 301: 135-149.

[38] Codignotto, J.O. and Aguirre, M. L. 1993. Coastal evolution in sea level and molluscan Fauna in northeastern Argentina during the Late Quaternary. Marine Geology, 110: 163-175.

[39] Cuerda, J., Vicens, D. and Gracia, F. 1991. Malacofauna y estratigrafía del Pleistoceno Superior marino de San Real (Santa Margalida, Mallorca). Bolletín de la Societat d'Història Natural de las Balears, 34; 99-108.

[40] Gowan, E. J., Rovere, A., Ryan, D. D., Richiano, S., Montes, A., Pappalardo, M., and Aguirre, M. L. (2021). Last interglacial (MIS 5e) sea-level proxies in southeastern South America (Earth Syst. Sci. Data, 13, 171-197.

[41] Kidwell, S. M. 2001. Preservation of species abundance in marine death assemblages. Science, 294: 1091-1094.

[42] Lario, J., Zazo, C., Goy, J. L., Somoza, L., Hoyos, M., Silva, P. G. and Hernández-Molina, F. J. 1993. Los episodios marinos cuaternarios de la costa de Málaga (España). Revista Sociedad Geológica de España, 6: 42-46.

[43] Martínez, S.; Ubilla, M., Verde, M., Perea, D., Rojas, A., Guéréquiz, R. and Piñeiro G. 2001. Paleoecology and Geochronology of Uruguayan Coastal Marine Pleistocene Deposits. Quaternary Research, 55: 246-254.

[44] Martínez, S., del Río, J. and Rojas, A. A. 2016. A Pleistocene (MIS 5e) mollusk Assemblages from Ezeiza (Buenos Aires Province, Argentina). Journal of South doi: 10.1016/j.jsames.2016.05.008

[45] Ortlieb, L.; Guzmán, N. and Candia, M. 1994. Moluscos litorales del Pleistoceno superior en el área de Antofagasta, Chile: Primeras determinaciones e indicaciones paleoceanográficas. Estudios Oceanológicos, 13: 57-63.

[46] Rojas, A. and Urteaga, D. 2011. Late Pleistocene and Holocene chitons (Mollusca, Polyplacophora) from Uruguay. Palaeobiogeography and palaeoenvironmental reconstruction in mid latitudes of the southwestern Atlantic. Geobios, 44: 377-386.

[47] Valentin, J. 1987. Noticia sobre un yacimiento de conchillas en el cementerio de Lomas de Zamora. Anales del Museo Nacional de Buenos Aires, 5: 227-231.

[48] Zazo, C., Goy, J. L., Hillaire-Marcel, C., González Delgado, J. A., Soler, V., Ghaleb, B, Dabrio, C., 2003b. Registros de los cambios del nivel del mar durante el Cuaternario en las Islas Canarias Occidentales (Tenerife y La Palma). Estudis geológicos, 59, 133-144.

[49] Cohen, K. M., Finney, S. and Gibbard, P. L. 2013. International Chronostratigraphic Chart. International Commission of Stratigraphy.

[50] Lutaenko, K. A. 1993. Climatic optimum during the Holocene and the distribution of warm-water mollusks in the Sea of Japan. Palaeogeography, Palaeoclimatology, Palaeoecology, 102: 273-281.

[51] Duplessy, E. -C., Ivanova, E., Murdmaa, E., Paterne, M. and Labeyrie, L. 2008. Holocene paleoceanography of the northern Barents Sea and variations of the northward heat transport by the Atlantic Ocean. Boreas, 30: 2-16. DOI:10.1111/j.1502-3885.2001.tb00984.x

[52] Funder, S. and Weidick, A. 1991. Holocene boreal mollusk in Greenland: palaeoceaonographic implications. Palaeogeography, Palaeoclimatology, Palaeoecology, 85 (1-2): 123-135.

[53] Hjort, C., Mangerud, J., Adrielsson, L., Bondevik S., Landvik, J. and Salvigsen, O. 1995. Radiocarbon dated common mussels *Mytilus edulis* from eastern Svalbard and the Holocene marine climatic optimum. Polar Research, 14 (2): 239-24

[54] Rohling, E. J. and De Rijk, S. 1999. The Holocene Climate Optimum and Last Glacial Maximum in the Mediterranean: the marine oxygen isotope record. Marine Geology, 153: 57-75.

[55] Salvigsen, O.; Forman, S. L. and Miller, G. H. 1992. Thennophilous molluscs on Svalbard during the Holocene and their paleoclimatic implications. Polar Research, 11(1): 1-10.

[56] Yuan, L.X.; Sun, L. G., Wei, G., Long, N., Xie, Z. and Yuhong, W. 2011. 9.400 yr B. P.: the mortality of mollusk shell (*Mya truncata*) at high Artic is associated with a sudden cooling event. Environmental Earth Sciences, 63: 1385-1393.

[57] Aguirre, M. L. 1990. Holocene Macrobenthic Molluscan Associations from North-eastern Buenos Aires Province, Argentina. Quaternary of South America and Antarctic Peninsula, 7: 161-195.

[58] Aguirre, M. L. 1993. Palaeobiogeography of the Holocene molluscan fauna from Northeastern Buenos Aires Province, Argentina: its relation to coastal evolution and sea

level changes. Palaeogeography, Palaeoclimatology, Palaeoecology, 102: 1-26.

[59] Aguirre, M. L. 2002. Óptimo climático en el Holoceno marino de la Argentina: evidencias Malacológicas. Actas XV Congreso Geológico Argentino, Calafate, Santa Cruz, Argentina. Tomo I, p. 548-553.

[60] Gaillard, M. J. and Lemdahl, G. 1994. Early-Holocene coastal environments and climate in southeast Sweden: a reconstruction based on macrofossils from submarine deposits. The Holocene, 4 (1): 53-68.

[61] Macsotay, O. and Cáceres, Hernández, R. C. 2005. Palaeoclimatology of the Pleistocene-Holocene using marine molluscs and hermatypic corals from northern Venezuela. Caribbean Journal of Earth Science, 39: 93-104.

[62] Parmesan, C. 2006. Ecological and Evolutionary Responses to Recent Climate Change. Annual Review of Ecology, Evolution and Systematic, 37: 637-674.

[63] Isla, F., Rutter, N., Schnack, E. and Zárate, M. 2000. La trasgresión Belgranense en Buenos Aires. Una revisión a cien años de su definición. Cuaternario y Ciencias Ambientales, 1: 3-14.

[64] Rostami, K., Peltier, W., Mangini, A. 2000. Quaternary marine terraces, sea – level changes and uplift history of Patagonia, Argentina: comparison with predictions of the ICE-4G (VM2) model of the global process of glacial Isostatic adjustment. Quaternary Science Reviews, 19: 1495-1525.

[65] Charó, M. P., Fucks, E. E., and Gordillo, S., 2015. Late Pleistocene-Recent marine malacological assemblages of the Colorado River delta (south of Buenos Aires Province): paleoecology and paleoclimatology. Quaternary International, 377: 52-71.

[66] Charó, M. P., Aceñolaza, G., Cavallotto, J.L. 2020. Ostrea stentina Payraudeau, 1826 found in a marine Deposit of Middle Late Pleistocene in the south of Buenos Aires Province, Argentina. I virtual Meeting of Systemtics, Biogeography δ Evolution (SBE). A joint Effort the Coronavirtual.

[67] Gonzalez, M. A., Weiler, N. E., Guida, N. C., 1986. Late Pleistocene transgresive deposits from 33°to 44°S, Argentine Republic. Journal of Coastal Research. Special Issues, N°1, 39'48.

[68] Fucks, E. E., Charó, M. and Pisano, F., 2012a. Aspectos estratigráficos y geomorfológicos del sector oriental patagónico bonaerense. Revista de la Sociedad Geológica de España, 25 (1-2): 29-44.

[69] Fucks, E. E. and Schnack, E. 2011. Evolución geomorfológica en el sector norte del golfo San Matías. XVIII Congreso Geológico Argentino, Neuquén, Argentina. Actas: 273-274.

[70] Weiler, N. E. 1988. Depósitos litorales del Pleistoceno tardío y Holoceno en Bahía Anegada, Provincia de Buenos Aires: Segunda Reunión Argentino de Sedimentología, Buenos Aires, Argentina. Actas, p. 245-249.

[71] Schnack, E. J.; Isla, F. I., De Francesco, F. O. and Fucks, E. E. 2005. Estratigrafía del Cuaternario Marino Tardío en la Provincia de Buenos Aires. Geología y Recursos Minerales de la Provincia de Buenos Aires. Relatorio del XVI Congreso Geológico Argentino, La Plata, Argentina. Capítulo X, p. 159-182.

[72] Fucks, E. E., Schnack, E. J. and Charó, M. P., 2012b. Aspectos geológicos y geomorfológicos del sector N del Golfo San Matías, Río Negro, Argentina. Revista de la Sociedad Geológica de España, 25 (1-2): 95-105.

[73] Angulo, R., Fidalgo, F., Gómez Peral, M. A. and Schnack, E. J. 1981. Geología y geomorfología del bajo de San Antonio y alrededores, provincia de Río Negro. Centro de Investigaciones científicas, Secretaria de planeamiento. Estudios y documentos, 8.

[74] Weiler, N. E. 1984. Rasgos geomorfológicos evolutivos del sector costero comprendido entre bahía Verde e isla Gaviota, provincia de Buenos Aires. Asociación Geológica Argentina, Revista XXXVIII (3-4): 392-404.

[75] Weiler, N. E. 2000. Evolución de los depósitos litorales en Bahía Anegada, Provincia de Buenos Aires, durante el Cuaternario tardío. Tesis Doctoral, Facultad de Ciencias Exactas y Naturales, Universidad de Buenos Aires, 184 p.

[76] Alberó, M. C., Angioloni, F. B, Codignotto, J.L., Linares, E., Weiler, N. E. 1980. Primeras edades Carbono-14 de afloramientos de conchillas de la Republica Argentina. Asociación Geológica Argentina. Revista XXXV(3): 363-374.

[77] Weiler, N.E. 1993. Niveles marinos del Pleistoceno tardío y Holoceno en Bahía Anegada, Provincia de Buenos Aires. Geocronología y correlación. Revista de la Asociación Geológica Argentina, 48 (3-4): 207-216.

[78] González, M.A. and Weiler, N. E. 1983. Ciclicidad de niveles marinos holocénicos en Bahía Blanca y en el delta del río Colorado. Simposio Oscilaciones de niveles del mar durante el Último Hemiciclo Deglacial en la Argentina. Revista de la Asociación Geológica, Mar del Plata, Actas 69-90.

[79] Trebino, L. 1987. Geomorfología y Evolución de las costas en los alrededores del pueblo de San Blas, Provincia de Buenos Aires. Revista de la Asociación Geológica Argentina, 42 (1-2): 9-22.

[80] Rutter, N., Schnack, E., Del Río, J., Fasano, J., Isla, F., Radkte, U. 1989. Quaternary litoral zones along the Patagonian coast, Argentina. Quaternary Science Reviews, 8: 213-234.

[81] Rutter, N., Radkte, U., Schnack, E. 1990. Comparison of ESR and amino acid data in correlating and dating quaternary shorelines along the Patagonian coast, Argentina. Journal of Coastal Research, 6 (2): 391-411.

[82] Bayer, S. and Gordillo, S., 2013. A new Pleistocene species of *Glycymeris* (Bivalvia, Glycymerididae) from northern Patagonia, Argentina. Ameghiniana, 50 (2): 265-268.

[83] Efremov, J. A. 1940. Taphonomy: new branch of paleontology. PanAmerican Geologist, 74: 81-93.

[84] Kidwell, S. M. 2002. Time-averaged molluscan death assemblages. Palimpsests of richness, snapshots of abundance. Geology, 30: 803-806.

[85] Kidwell, S. M. 2008. Ecological fidelity of open marine molluscan death assemblages: effects of post-mortem transportation, shelf health, and taphonomic inertia. Lethaia, 41: 199-217.

[86] Kidwell, S. M. 2013. Time-averaging and fidelity of modern death assemblages: building a taphonomic foundation for conversation palaeobiology. Palaeontology, 56: 487-522.

[87] Kidwell, S. M. and Tomasovych, A. 2013. Implications of Time-Averaged death assemblages for ecology and Conservation biology. Annual Review of Ecology, Evolution, and Systematics, 44: 539-563.

[88] Linné, C. 1758. Systema Naturae per regnatria naturae, secundum classes, ordines, genera, species, cum characteribus, differentiis, synonymis, locis. Edition decima, reformata [10th revised edition],vol. 1: 824.

[89] Adams, A. 1856. Descriptions of thirty four New species of bivalve Mollusca (Ledo, Nucula and Pythina) from the Cumingian Collection. Proceedings of the Zoological Society of London, 24: 47-53

[90] Sowerby G. B., I. 1833. Descriptions of new species of shells from the collection formed by Mr. Cuming on the western coast of South America, and among the islands of the southern Pacific Ocean. Proceedings of the Committee of Science and Correspondence of the Zoological Society of London. 2 ("1832"): 194-202.

[91] Molina, J. I., 1782. Saggio sulla storia naturale del Chile, del Signor Abate Giovanni Ignazio Molina. Bologna. i-v, 1-306.

[92] Lamarck J.B. P.A. 1819. Histoire naturelle des animaux sans vertèbres, Paris.Tome 6(1): 343 pp.

[93] Lamarck, J. B. 1801. Système des animaux sans vertèbres, ou tableau général des classes, des ordres et des genres de ces animaux; Présentant leurs caractères essentiels et leur distribution, d'apres la considération de leurs rapports naturels et de leur organisation, et suivant l'arrangement établi dans les galeries du Muséum d'Histoire Naturelle, parmi leurs dépouilles conservées; Précédé du discours d'ouverture du Cours de Zoologie, donné dans le Muséum National d'Histoire Naturelle l'an 8 de la République. Published by the author and Deterville, Paris: viii + 432 pp.

[94] Say, T., 1830-1834. American Conchology, or descriptions of the shells of North America. Illustrated by coloured figures from original drawings executed from nature. School Press, New Harmony, Part 6 (April 1834); Part 7 (1834?, published after Say's death; edited by T. A. Conrad).

[95] Guilding, I. 1828. Observations on the zoology of the Caribbean Islands. The Zoological Journal. 3: 527-544.

[96] Thunberg, C.P. 1793. Tekning och Beskrifning på en stor Ostronsort ifrån Japan. Kongliga Vetenskaps Academiens Nya Handlingar. 14(4-6): 140-142.

[97] Ihering, H.V. 1907. Les Mollusques fossiles du Tertiaire et du Cretace superieur de l'Argentine. Anales del Museo Nacional de Buenos Aires, Serie III, 1-611.

[98] Signorelli, J. and Scarabino, F. 2010. Mactra guidoi n sp. and Mactra patagonica (Bivalvia: Mactridae) two long misunderstood species from southwestern Atlantic Ocean Malacologia, 52 (1): 31-52.

[99] Lamarck, J. B. P. A. 1818. Histoire Naturelle des Animaux Sans Vértebrés. Paris Tome 5, 612 pp.

[100] Reeve, L. A. 1854. Monograph of the genus Mesodesma. In: Conchologia Iconica, or, illustrations of the shells of molluscous animals, vol. 8. L. Reeve & Co., London. Pls 1-4.

[101] Hanley S.C.T. 1842-1843. Illustrated, enlarged and English edition of Lamarck's species of shells forming the third edition of the Index Testaceologicus. London: Wood. 224+8+[3]+8 pp., 3 pls. [pp. 1-32. pl. 1-2. late 1842; pp. 11-32 [reissue]

[102] Smith, E. A. 1885. Report on the Lamellibranchiata collected by HMS Challenger during the years 1873-76. Report on the Scientific Results of the Voyage of H.M.S.Challenger during the years 1873-76, Zoology 13: 1-341, pls 1-25.

[103] Lightfoot, J. 1786. A Catalogue of the Portland Museum, lately the property of the Dutchess Dowager of Portland, deceased; which will be sold by auction by Mr. Skinner & Co. [book]. London. viii + 194 pp.

[104] Gmelin, J. F. 1791. Vermes. In: Gmelim J. F. (Ed.) Caroli a Linnaei Systema Naturae per Regna Tria

Naturae, Ed. 13. Tome 1 (6). G. E. Beer, Lipsiae [Leipzig]. pp. 3021-3910.

[105] Dillwyn, L. W. 1817. A descriptive catalogue of Recent shells, arranged according to the synonymy London: John and Arthur Arch, Vol 1: 1-158. Vol 2: 581-1092.

[106] King P.P. 1832. Description of the Cirrhipeda, Conchifera and Mollusca, in a collection formed by the officers of H.M.S. Adventure and Beagle employed between the years 1826 and 1830 in surveying the southern coasts of South America, including the Straits of Magalhaens and the coast of Tierra del Fuego. Zoological Journal, 5: 332-349.

[107] Pilsbry, H. A. 1897. New species of mollusks from Uruguay. Proceedings of the Academy of Natural Sciences of Philadelphia. 49: 290-298.

[108] Couthowy, J.P. 1839. Monograph on the family With remarks on two species of Patelloidea, and descriptions of new Species of marine shells, a species of Anculotus and one of Eolis. Boston Journal of Natural History 2 (2): 129-189.

[109] Martens, E. von. 1900. Neue Fissurella aus Südbrasilien. Nachrichtsblatt der deutschen malakozoologischen Gesellschaft. 32(11/12): 187.

[110] Smith, E. A. 1880. Descriptions of five new species of shells from Uruguay. Annals and Magazine of Natural History. ser. 5, 6 (34): 319-322.

[111] Collin, R. 2005. Development, Phylogeny and taxonomy of Bostry-Capulus (Caenogastropoda Calyphacidae), an ancient cryptic radiation. Zoological Journal of the Linnean Society, 144:75-101.

[112] Simone, L. R., Pastorino, G., Penchaszadeh, P. E. 2000. Crepidula argentina (Gastropoda: Calyptraeidae) a new species from the littoral of Argentina. The nautilus, 114 (4): 127-141.

[113] Lamarck, J.B. M. de. 1822. Histoire naturelle des animaux sans vertèbres. Tome septième. Paris: published by the Author, 711 pp.

[114] Pallas, P.S. 1774. Spicilegia Zoologica quibus novae. Imprimis et Obscurae Animalium Species. Part 10. Berolini, Lange, 267 pp.

[115] Donovan, E. E. 1823-1827. The Naturalist's repository, or monthly miscellany of exotic natural history, etc. [Book series]. In 5 volumes 1 (1823)-5 (1827).

[116] Lamarck, J.B. 1811. Suite de la détermination des espèces de Mollusques testacés. Annales du Muséum National d'Histoire Naturelle. 16: 300-328.

[117] Broderip, W.J. 1836. Description of some Species of shells apparently not hitherlo Recorded. Proceedings of the Zoological. Society of London, 4: 43-45.

[118] Duclos, P. L. 1835-1840. Histoire naturelle générale et particulière de tous les genres de coquilles univalves marines a l'état vivant et fossile publiée par monographie. Genre Olive. Paris: Institut de France. 33 plates: pls 1-12

[119] Klappenbach M.A. 1965. Consideraciones sobre el género Olivancillaria d'Orbigny 1840 (Moll. Gastr.) y descripción de dos nuevas especies de aguas Argentinas y Uruguayas. Com.Zool.Mus.His.Nat. Montevideo, 8:104.

[120] Kiener L.C. 1834-1841. Spécies général et iconographie des coquilles vivantes. Vol. 9. Famille des Purpurifères. Deuxième partie. Genres Buccin (Buccinum), Adanson, pp. 1-112 + table with duplicate page numbers 105-108, pl. 1-31 [pp. 1-64 (1834),

65-104 and 105-108 of table (1835), 105-112 of text (1841); pl. 1-24 (1834).

[121] Doello-Jurado, M. 1938. Nuevos datos sobre la fauna marina de la meseta continental de la Argentina y del Uruguay. Phys., 12: 279-292.

[122] Castellano, Z. 1982. Los Pyramidellidae de la República Argentina (Mall, Entomo taeniata). Comunicaciones del Museo Argentino de Ciencias Naturales ¨Bernardino Rivadavia¨. Hidrobiología 2(7): 61-85.

[123] Blainville, H.M. D. de 1824. Mollusques, Mollusca (Malacoz.). In: Diccionnarie des Sciences Naturelles (F. Cuvier, ed.) Vol 32: 1-392. Leuralt, Strasbourg et Paris, δ Le Normant, Paris.

[124] Aguirre, M. L. and Fucks, E. E., 2004. Moluscos y paleoambientes del Cuaternario marino en el sur de Entre Rios y litoral bonaerense. Temas de Biodiversidad del litoral fluvial argentino. Miscelánea, 12: 55-10.

[125] Tonni, E. P. and Fidalgo, F. 1978. Consideraciones sobre los cambios climáticos durante el Pleistoceno Tardío–Reciente en la provincia de Buenos Aires. Aspectos ecológicos y zoogeográficos relacionados. Revista de Asociación Paleontológica Argentina. Tomo XV (1-2): 235-253.

[126] Arruda, C. C. B., Beasley, C. R., Vallinoto, M., Marques-Silva, N. S. and Tagliaro, C. H. 2009. Significant genetic differentiation among populations of *Anomalocardia brasiliana* (Gmelin, 1791): a bivalve with planktonic larval dispersion. Genetic Molecular Biology, 32: 423-430.

[127] Oliveira, I.; Amorim, A., Lavander, H., Peixoto, S. and Gálvez, A. O. 2011. Spatial and temporal distribution of the shellfish *Anomalocardia brasiliana* (Gmelin, 1791) on Mangue Seco beach, Pernambuco, Brazil. International Journal of Acuatic Science, 2: 68-79.

[128] Martínez, S.; Rojas, A., Ubilla, M., Verde, M., Perea, D., Piñeiro, G. 2006. Molluscan assemblages from the marine Holocene of Uruguay: composition, geochronology, and paleoenvironmental signals. Ameghiniana, 43 (2): 385-397.

[129] Walker, M.; Johnsen, S., Rasmussen, S. O., Popp, T., Steffensen, J.P., Gibbard, P., Hoek, W., Lowe, J., Andrews, J., Björck, S., Cwynar, L.C., Hughen, K., Kershaw, P., Kromer, B., Litt, T., Lowe, D.J., Nakagawa, T., Newnham, R. and Schwander, J. 2009. Formal definition and dating of the GSSP (Global Stratotype Section and Point) for the base of the Holocene using the Greenland NGRIP ice core, and selected auxiliary records. Journal of Quaternary Science, 24(1): 3-17.

[130] Veliz, D. and Vásquez, J. A. 2000. La Familia Trochidae (Mollusca: Gastropoda) en el norte de Chile: consideraciones ecológicas y taxonómicas. Revista Chilena de Historia Natural, 73: 757-769.

[131] Díaz, A. and Ortlieb, L. 1993. El fenómeno del Niño y los moluscos de la costa peruana. Bulletin de Institut Francais d'Estudes Andines, 22 (1):159-177.

[132] Ortlieb, L. and Díaz, A. 1991. Distribuciones de moluscos litorales del Perú en el Pleistoceno Superior: primeras interpretaciones paleoceanográficas y paleoclimáticas. IIIa Reunión anual Proyecto PICG, 281, p. 39-56.

[133] Jackson, D. S., Méndez, M. C., López, M. P. and Sequel, R. Q. 2005. Evaluación de un asentamiento arqueológico en el semiárido de Chile: procesos de formación, fauna extinta y componentes culturales. Intersecciones en Antropología, 6: 139-151.

[134] Baez, P. R. and Jackson, D. S., 2008. Exploitation of loco, *Concholepas*

concholepas (Gastropoda: Muricidae), during the Holocene of Norte Semiárido, Chile. Early Human Impact on Megamolluscs Publishers of British Archaeological Reports, 79-94 p.

[135] Cárdenas, J. and Gordillo, S. 2009. Paleoenviromental interpretation of late Quaternary molluscan assemblages from southern South America: A taphonomic comparison between the Strait of Magellan and the Beagle Channel. Andean Geology, 36(1): 81-93

[136] Moreno, C. A. 2004. Efectos de El Niño en el reclutamiento de *Concholepas concholepas* y *Tegula atra* (Mollusca, Gastropoda) en la costa de Valdivia, Sur de Chile. En Avaria y otros Eds. El Niño-La Niña 1997-2000 Sus efectos en Chile. Comité Oceanográfico Nacional (CONA), p. 179-190.